# Information, Economics and Power

## The North-South Dimension

*edited by* Rita Cruise O'Brien

HODDER AND STOUGHTON
LONDON SYDNEY AUCKLAND TORONTO

**British Library Cataloguing in Publication Data**
Information, economics and power,
1. Underdeveloped areas – Foreign economic relations
I. O'Brien, Rita Cruise
382.1 HF1428

ISBN 0 340 33614 5

First published 1983



Represented in Nigeria and Cameroon by Nigeria Publishers Services Ltd, P.O. Box 62, Ibadan, Nigeria

Represented in East Africa by K. W. Martin, P.O. Box 30583, Nairobi, Kenya

**Printed in Great Britain for Hodder and Stoughton Educational, a division of Hodder and Stoughton Ltd, Mill Road, Dunton Green, Sevenoaks, Kent, by Biddles Ltd, Guildford, Surrey.**
**Photoset by Rowland Phototypesetting Ltd, Bury St Edmunds, Suffolk.**

# Contents

# Preface

This book considers how differential access to information may affect the outcome of international negotiations, which are of great importance to developing countries. In contrast to the debate on free flow of information which attracted headlines in the 1970s, it does not deal with the mass media. It looks rather at a concrete set of issues and negotiating strategies in several types of bargaining situations internationally, which are directly or indirectly affected by information problems. It considers also how new technology in information and telecommunications may add a new dimension to North-South inequality in the years to come.

In the autumn of 1980 Gerry Helleiner and I published an article in *International Organisation*, an American scholarly journal. A revised version of that article is included in this volume. We called for thinking and research on the problems of information access and the impact of new technology on traditional areas of North-South bargaining. We felt that if the South did not consider these problems promptly, it risked falling further and further behind in the aim to achieve a better deal with multinationals and the governments of industrialised countries. A modest agenda of items for research was thus outlined.

The next stage was, we felt, to offer for scrutiny the problems as we saw them, by assembling a group of specialists on different sectors of North-South negotiation to discuss the problems we had outlined.

The Development Centre of the OECD provided sponsorship for a conference on the subject at the Institute of Development Studies. The conference, 'Information Access and International Negotiations: New Directions in North-South Relations', took place in Sussex from 29 September to 2 October 1981. Three questions were provided to participants by way of orienting conference discussions:

- What was the impact of *the information factor* in your particular area of international negotiation?
- What effect will *the information revolution* have on those negotiations in which the developing world has much at stake?
- How might the most effective *intelligence units* be designed to serve the needs of developing countries to try to redress current problems?

The chapters which make up this volume are largely the revised and edited versions of the papers presented to that conference by the authors. There are two important additions: William Lazarus on remote sensing and mineral negotiations, and Björn Tell on the organisation of information capability in developing countries. Other participants in the conference included Carlos Diaz-Alejandro, Professor of Economics at Yale University; Hans Gassmann, Director of the Information, Computer and Communication Policy group at the OECD; Raymond Aubrac of the International Development Research Council of Canada, who designed AGRIS, the information system of the UN Food and Agriculture Organisation; Peter Lai, the Secretary General of the International Tin Council (formerly with the International Rubber Council); and my colleagues at IDS: Philip Daniel, former adviser to the Ministry of Mines in Papua New Guinea, and Reginald Green, former chief advisor to the Treasury in Tanzania. The proceedings were much enriched by their contributions. I would like to thank most warmly my secretary at the Institute, Jennifer Kearns, who prepared this manuscript for publication.

If I may be allowed, in closing, to go back to the origins of this book, I wish to give some background to the emergence of these ideas in my own thinking. From 1973 to 1979 I worked principally on the international influences on the development of mass media systems in the Third World. In that period I naturally got caught in the cross-fire of the raging debate in UNESCO and elsewhere on the free flow of information. As a researcher working in an industrialised country on development problems, I felt in sympathy with neither the ideas expressed by the Ministers of Information of developing countries nor those expressed by the spokesmen of the Western press. Somewhat war weary, I wanted to duck out of the ideological debate in order to apply my knowledge and skills to something which could help the interests of the South, while working still in Europe. To the people, who have participated directly and indirectly in helping me to begin this task, some of whom are mentioned above, and the remainder of whom are found in the Table of Contents, I wish to offer my gratitude for a rich and rewarding education. But I must thank most particularly Gerry Helleiner, whose characteristic prescience in recognising the importance of information in the international debate, and widely admired command of international and financial problems, has deeply enriched the learning process. His wisdom, good judgement and friendship were of great significance in the completion of this task.

Rita Cruise O'Brien
January 1983

# List of Contributors

*Dr Rita Cruise O'Brien* is a Fellow at the Institute of Development Studies, University of Sussex. She has recently completed work for the OECD on information flows to developing countries, and is currently planning a book, *The Politics of the Information Age.*

*Professor G. K. Helleiner* is Professor of Economics at the University of Toronto. He has worked as a consultant to UNCTAD, ILO and the World Bank. His most recent books are *International Economic Disorder: Essays in North-South Relations*, 1981, and *Intra-Firm Trade and the Developing Countries*, 1981.

*Dr Karl P. Sauvant* represented the Centre on Transnational Corporations at the United Nations in New York. His most recent publications are *The Group of 77: Evolution, Structure, Organisation*, 1981, and *Changing Priorities on the International Agenda: The New International Economic Order*, 1981.

*Dr Stephany Griffith-Jones* is a Research Fellow at the Institute of Development Studies. She has just published a book, *The Role of Finance in the Transition to Socialism*, 1982.

*Dr Roberto Junguito* is the President of the Colombian Agricultural Society.

*Dr Diego Pizano* is Head of Economic Research of the Colombian Federation of Coffee Growers.

*Mr Carlos Fortin* is the Deputy Director of the Institute of Development Studies. From 1971–73, he was the Representative of Chile to the Intergovernmental Council of Copper Exporting Countries (CIPEC). His latest publication is 'The Mining Sector', Technical Paper No. 16 in *Basic Needs in an Economy under Pressure*, Findings and Recommendations of the ILO/JASPA Basic Needs Mission to Zambia, 1981.

*Mr William Lazarus* is doing research in the research programme on Communications Policy at the Massachusetts Institute of Technology. He was formerly Asst Station Manager and Director of

Special Projects at the Educational Television Center, Menlo Park, California.

*Dr Nicholas Jéquier* is Professor at the Institut de Hauts Études en Administration Publique, University of Lausanne. He was until recently at the Development Centre of the OECD, where he worked principally on technology transfer problems. His latest book is *The World of Appropriate Technology*, 1981.

*Dr Björn V. Tell* is Director of Libraries at the University Library of Lund in Sweden. As rapporteur of an OECD *ad hoc* High-level Group he was responsible for the book, *Information for a Changing Society – Some Policy Considerations* (Paris, OECD, 1971). For UNESCO he has also published *Design and Planning of National Information Systems.* (Paris, UNESCO, 1976).

*Professor Miguel Wionczek* is Senior Research Fellow and Director of Long-Range Energy Research Program at El Colegio de México. His most recent book *Some Key Issues for the World Periphery* has just appeared at Pergamon Press.

# Introduction

By the beginning of the 1980s, North-South problems had been recognised as a major area of international concern. The publication of the first Brandt Report* gave development problems particular international significance. The Report recommended new forms of global interdependence in order to promote development in the poorer regions of the world. Its 'Agenda for action' included increased financial recycling to developing countries, greater access to rich country markets for manufactured goods from newly industrialising countries, and better prices for commodities. Its succinct political message was the urgent need to develop a more equitable sharing of global resources. In the past three years concrete achievements on the Agenda are few, and the latest World Development report of the World Bank emphasises again how little progress has been made. The second report of the Commission in 1983** draws attention to the common crisis which affects us all. Severely declining growth rates in Latin America and total economic dislocation in Africa have generated some serious debate about the constraints on immediate relief and longer-term structural assistance by the World Bank and International Monetary Fund. Repeated difficulties for producer countries in establishing commodity agreements have recently promoted new consideration of collective self-reliance in the Third World. Strengthening the aims of the South requires immediate concrete action, beyond the rhetoric of the 1970s. This book is designed to address itself to a specific set of issues which will hopefully provide a modest step towards the achievement of those aims.

Some argue that information is plentiful. The concept of free trade, for example, rests on a doubtful supposition of free access to information. There are in reality dramatic imperfections in information markets. As Kenneth Arrow has written, 'The presumption that free markets will lead to an efficient allocation of resources is not valid . . . There are at least two salient characteristics of information which prevent it from being fully identified as one of the commodities represented in our abstract models of equilibrium: it is by definition indivisible in its use and very difficult to appropriate'.[1] This is of considerable significance since differential capacity in the acquisition and use of

* Report of the Independent Commission on International Development Issues, 1980.
** Common Crisis: North-South Cooperation for World Recovery.

specialised information in commercial and intergovernmental negotiations directly affects the bargaining capability of developing countries. This phenomenon, which often yields instability in the negotiating process, is in the interests of neither North nor South. Negotiation has long since ceased being a matter of guile and craft; it resembles much more the management of complex sources of information, and risk assessment.

Before one considers the imperfections in information markets and their effect on negotiation, an important recent trend must be acknowledged. New information and telecommunication systems are dramatically altering the whole nature of international business. The extension of modern management, marketing and corporate planning is based on these systems: decision and communication has been fundamentally transformed by the new technology.[2] This adds a new dimension to all types of international negotiation with transnational corporations, which has hitherto been little considered. Are those countries with scarce resources and few assets in negotiation going to fall further and further behind by the end of the decade? Whatever political progress may be made or price levels maintained, these new systems may in the short-run seriously exacerbate the information capacity differential between North and South.

A crucial question for developing countries is how power over resources may be converted into power over events. In certain instances government or parastatal representatives may not be able to make effective use of the resources under their control in order to bargain effectively. This has serious consequences because of the high cost of establishing alternative markets and sources of supply. There is a large gap between the aims of restructuring the international economy and achievements in specific areas. Some would say that the problems are more political than informational; a facet of the incapacity to create an institutional base to serve the needs of the South on a continuous basis, or the political will to promote genuine regional or sectoral cooperation among developing countries.

To date the broad aims of the New International Economic Order may have put incremental strategies at a disadvantage. Given the current lack of progress on the broad front, strategies for incremental improvements ought to be the prime target for research and action. Discussions during the last five years on the formation of a secretariat to serve the needs of the South have met repeated impasse – largely for political reasons. Yet without a Bank for International Settlements or an OECD to serve its interests, the South is inevitably weakened further. For the present, organisations such as UNCTAD, UNIDO and other UN specialised agencies operate surrogate secretariats on behalf of the Group of 77. But in the long run, this can only be a temporary solution to some of the problems of information access.

Large multilateral information systems and data banks designed for the use of developing countries (DEVSIS, UNISIST, AGRIS or the Development Information Network of the UNDP) all suffer from a number of problems which leave a large gap in specific user-designed information requirements. Their under-utilisation by the countries which most need them, and their research or project orientation in contrast to immediate negotiating requirements, leave much scope to the private consultancy market. There are gross inefficiencies in the provision of expertise in trade and financial negotiations. To have the basic information on where to look for information, or how to 'plug into the expertise network', can be of greatest value in certain instances.

Not all negotiations have an identifiable or even plausible information gap problem. Many problems are attributable to political issues or intractable market situations. As the most likely progress to be achieved by the South in the 1980s must be deliberate and painstaking in its efforts to promote movement in international economic negotiations – often chipping away at the margins of possible new advantage – information becomes a crucial input.

## New information technology: a mixed blessing for developing countries

New information and telecommunication technologies facilitate our capacity to reshape world trade, international finance, agricultural production and food security, and to cope more effectively with natural and other crises such as extreme weather conditions, crop failures, and epidemics. The same systems, however, offer the prospect of increased centralisation and power in industrialised countries and their main business centres. There is an inherent paradox in the dual tendency towards global technical integration and political disintegration in a climate of reduced growth in almost all sectors of the economy. As the tendency for greater interdependence may be accompanied by growing fragmentation, malfunctioning of the systems is all the more unsettling, if mutual confidence is lacking. The 'communications explosion' holds the potential of both threat and promise for the developing nations.

Information and telecommunication problems in themselves are beginning to emerge as an important North-South issue. The developing world is concerned about its data dependence, the control of data across national boundaries, the allocation of space and the spectrum, and the enlarged potential of remote sensing satellites for altering the stakes of international bargaining. For their part, transnational corporations support 'business as usual' in the face of demands for a 'new information order'.

A key consideration in the negotiating power of developing countries is the extent of information available to detect and evaluate the practices of transnational corporations with which many negotiations are transacted. New trends in 'the information economy' do not stop at national boundaries; they are most importantly contained in the new information systems of these corporations. The Centre on Transnational Corporations at the UN in New York has begun to seriously consider these problems: the contribution by a member of the Centre's staff is one of the early papers in their programme. It considers the role of international business in commercial and corporate transborder data flows, documenting the principal features of this new phenomenon and outlining its potential benefits and problems. Above all, an awareness of the issues created by international data traffic and their effect on developing countries is of central significance. There will be increased activity in the assessment and possible regulation of international data traffic in the coming years.

The Ministerial meeting of the General Agreement on Tariffs and Trade (GATT) in November 1982 discussed major changes in service and high technology sectors: the effort to pre-empt the consideration of trade barriers in data has compelled the Americans to make it an important agenda item. Not previously encountered as a trade issue, data traffic raises some specific problems in terms of definition, identification, enumeration and detection: it is physically impossible to impose upon data the same kinds of regulations imposed on goods or printed information. The spectre of control is of vital policy importance to business firms. It reflects the fact that technological progress has altered in significant qualitative terms the nature of international production, trade and finance.

National governments in Europe began to introduce omnibus legislation on privacy protection in the early 1970s, designed to prevent the transmission of personal data across national boundaries.

For many countries the importance and impact of new technology and data traffic goes beyond privacy to issues of trade and economic development. Several newly industrialising countries, including Brazil and Mexico, have recently developed comprehensive policies on data processing and transmission. Transnational banks and carriers like American Express have reacted sharply to this trend, which they view as a threat to the climate of private investment in developing regions. We are thus in a new round of debate on national sovereignty and transnational business, which has been with us for a long time, but with a new added technological dimension, raising a new and complicated set of challenges.

## Building negotiating capacity

It is the application of knowledge and organisation, skill and planning in the use of information, whether produced or transmitted by ordinary or electronic means, which may be the most crucial factor in negotiating advantage: the capacity to synthesise information is itself based on innovation and is vital to negotiating power. Thus, while keeping a watching brief on changes in new high technology systems and new areas of negotiation generated by them, the most immediate problems of differential access to information in the negotiating environment and what might be done to improve the situation concern the acquisition and management of information at a non-technical level. Much of our conference discussions dealt with the issues at this level.

The economics literature on information and markets, which is considered by Gerry Helleiner in our joint contribution to this volume, was a starting point of conference discussions. Consideration and discussion of the literature concluded that it was not particularly helpful in trying to understand the problems of risk in North-South negotiations. In his own contribution on information uncertainty Professor Helleiner makes a number of practical observations and suggestions about information acquisition and consumer protection to deal with the market problems currently encountered by negotiations from the South. He lists various devices for altering the distribution of contract risk. The emphasis of discussion on the general problem sought to guard against the inevitable mystification surrounding risk assessment by means of econometric modelling and forecasting. Such techniques have often been part of the tactics used by advanced industrialised countries, in commodity and mineral negotiations. This included the use of econometric models – for example in what seemed like a deliberate attempt to 'blind with science' but without really crucial significance to the outcome. Efforts to develop or use such techniques by the negotiating terms of developing countries were felt to be much less policy effective than a consideration of the knowledge, skill and information synthesis inputs to building local capability on financial markets, commodities futures or economic forecasting.

Information on borrowing and financial management has been an essential requirement to many of the more prosperous developing countries in recent years. But as Stephany Griffith-Jones outlines in her contribution, very few countries indeed account for more than 80% of developing country borrowing. The remainder are newcomers to Euro-bond and Euro-dollar markets and have much to learn. The international credit market has information limitations on the side of borrower and lender. Thus, financial consulting services have become big business and, in addition to some of the large commercial banks and consortia, private consultants have become very powerful purveyors of information. In consideration of a rationalisation of these services, suggestions for information sharing, technical cooperation arrangements or use of regional development banks were considered in the discussion on financial services. Frequent borrowers, it was noted, can become very proficient at 'operating the system' – borrowing at low spreads, for example, thus providing a source of advisory capacity which may be used by other borrowers. At a much more fundamental level is the distribution of power and financial control in the world's major intergovernmental institutions such as the World Bank and the International Monetary Fund. With little individual power and paultry collaborative capacity among themselves, the South currently commands little influence in these institutions. Dr Griffith-Jones emphasises the political significance of this situation and the information components contained within a range of financial problems.

In commodity negotiations, conference discussions favoured the view that while information problems may not ultimately be crucial to the outcome of negotiation, they are certainly of significance. Peter Lai, who has been closely involved with the attempts to establish cocoa, rubber and tin agreements at the international level, emphasised that political rather than informational factors were the ultimate stumbling block to achieving agreements on a more equitable distribution of benefits. In the contributions on coffee and copper in this volume, the authors analyse the scope and nature of information problems in recent negotiations. Political difficulties, and lack of mutual confidence among producer nations lie behind the limited successes of producer associations like CIPEC, in copper, or the short-lived coffee association, the Bogota Group. Some technical, market and financial information has been provided by these organisations. The analysis of the information capability creation contained in the chapters on coffee and copper provides a lesson for how such facilities might have improved and extended to other sectors.

The use of remote sensing satellites for the detection of mineral deposits or agricultural output, and the use of satellite data for altering the stakes of negotiation were raised in our discussions. In order to clarify certain suppositions about the use of this data in mineral negotiations, it was a great asset to be able to include William Lazarus' paper to complement the discussion. Among the new problem areas for developing countries generated by high technology, remote sensing devices and differential capability in the use of its processed data must take prominence. Remote sensing may be seen as both a threat and promise. The Lazarus chapter spells out in detail just what is behind the emerging debate on the use of satellite data in negotiation, the impact of differential access to data and what might be done about it. For the moment a number of experts who advise mineral producing countries in the developing world do not feel it is essential to invest in significant processing capability for remote sensing data. Yet it is significant to note that the geologists of major mining corporations, in collaboration with the American Geosat Committee, a service organisation, use data as a means of checking calculations and estimations of prospective mineral deposits. The high cost of prospecting and the uncertainty about ownership contracts with host country governments might make the use of this type of data more pertinent in the years to come.

## Consultancy market and data banks

For many developing countries with high level manpower deficiencies, the capacity to organise information in terms which are appropriate in respect of speed and specificity may be most effectively achieved in the short-run by buying-in outside expertise. At present this is available to developing countries seeking to improve their negotiating position either through UN agencies or other forms of bilateral aid, or through purchases on the market – consulting firms in particular. Unless consultants identify with the policy aims of the country on behalf of which they are negotiating, their interests are purely commercial. In this case local collaborators need to know how to frame questions in order to derive the best type of information appropriate to their purposes. This requires considerable knowledge of the parameters and substance of negotiations. The knowledge of a consultant is naturally 'embodied information' in the sense that it is usually used in a one-off situation and not necessarily left to be used or built upon subsequently, unless local capability is developed.

For many countries knowledge of the network or how to 'plug in to it' is a most rudimentary requirement. Private consultants charge very high rates and less than adequate use of their expertise can be very costly. The UNIDO Joint Study Group on International Industrial Co-operation recommended in 1979 a proposal to supplement expertise available only in the private market by the formation of a Consultancy Agency, financed on a downward sliding scale by the UN initially and later through voluntary contributions. Nothing to date has come of this. An immediate practical consideration might be a directory of specialised consultants in different fields of negotiation, a register might be updated and verified by relevant inter-governmental agencies.

Courses on negotiation have been offered by various UN specialised agencies and other bodies, although the most valuable learning experience for negotiation is certainly learning 'on the job' as part of a team. There is great strength in trying to keep the local team together following negotiation for the purposes of monitoring the agreement, the general policies of the company, and for obliging the firm to log copies of surveys and reports with the relevant parts of the government. The perennial problem of mobility among senior civil servants in many developing countries often makes this task very difficult.

Among recent institutional initiatives made on behalf of developing countries is the creation of data banks to serve their needs. The most important problem to be found in all these large schemes is their under-utilisation, raising questions therefore about the considerable technical, financial and expert resources which they require, and questioning their future viability, no matter how well-conceived or designed they may be. In comparative terms, they are less utilised by the poorer countries, where information needs are perhaps the most dramatic. Although considerable effort was made to solicit views on system-design, the very nature of some of these large data-base systems engenders an inevitable designer-bias.

The capacity to use information systems effectively and efficiently is a sophisticated working method of advanced industrialised societies. Just as a user requires some basic information to know how to obtain or where to find more information, so the definition of data requirements in relation to a system may prove to be a great obstacle. In general, users are singularly unimaginative in expressing a need for information services with which they are not familiar. Similarly, it might be added, a potential user in a developing country may have great difficulty in trying to transform the need for information he/she perceives into something to which an information system can respond. Large scale, long-range systems planning may lead to large-scale failure, which would be costly for developing countries with limited national income. It might be as well instead to foster evolutionary mechanisms that encourage innovation. Matching the information needs of developing countries with the provision of what is required inevitably goes through the vicissitudes of a very complex process, to which those who favour the large system approach must be continually mindful.

## Building national capability

Improving the intelligence capability of developing countries is of prime concern in helping to redress differential access to information. How this might be achieved was among the most concrete features of our conference deliberations. There are certain types of information, knowledge and skills which are best maintained at the local level in order to avoid the difficulties of obtaining market intelligence using effectively data banks. An OECD Conference in 1980 introduced the concept of social intelligence for development and raised questions about the use of existing systems and creation of new ones, tailor-made for policy problems (proceedings forthcoming as *The Knowledge Industry and the Process of Development, 1983*).

Building an intelligence capability can be a long-term subscription process which may be short cut by careful recruitment, organisation and task design. Local capacity to filter irrelevant information and avoid information overload requires a skilful definition of policy aims. In intelligence, there is a knowledge of ignorance and a means of buying in from the market whatever resources are essential. In the development of intelligence capacity for energy planning, Miguel Wionczek outlined the types of periodical and other documentation essential to the task. His contribution to this volume gives some practical suggestions for avoiding wastage of effort and using personal and financial resources to optimal capacity. He outlined the type of professional most suited to intelligence work, and the design of a team. Knowledge capability for national intelligence cannot be transferred, like discreet bits of data or packaged information: it is rather something which must be developed in the local context in relation to specific policy problems. The wholesale transfer of academic and professional structures from industrialised countries to the Third World has created work orientation, priorities and institutions, which have not necessarily served local needs best. One of the aims therefore in the creation of a national intelligence capability is to overcome some of the more penetrating aspects of dependence, which formal educational and accreditation structures have helped to sustain in developing countries. High motivation, an interdisciplinary background and a capacity for synthesis of information (or filtering it for specific purposes to make it usable) are among the characteristics recommended for intelligence team personnel. Their task is the transformation of information, received from diverse sources, into active and usable knowledge.

The location of such an information unit or intelligence group in the national management structure was another aspect of conference discussion. A prime consideration was that these units maintain an autonomous relationship to the central administration and thus preserve their independence, that they have an agility of action, and a specialised staff with qualifications and aptitude for dealing with intelligence problems. In his contribution Nicholas Jéquier gives a broad overview of some of the basic characteristics of this process, and goes on to consider investment in various forms of communication for development purposes. Building a technological intelligence capability was one of the priority areas of consideration, experience from which may be derived from studying such examples as the Korea or Singapore experience. Having been instrumental in the organisation of the OECD conference which preceded ours, Professor Jéquier is one of the early advocates of the intelligence approach to development. The chapter by Björn Tell complements this approach to the problem by describing in concrete detail a national model of information management in a newly industrialising country. In this case, a Ministry of Mining and Energy established an intelligence function to cater for technology transfer: information network and administrative linkages are described in some detail. It is obvious that the economy of location in the user country was of prime importance.

## Conclusion

The need for collaboration among developing countries, including the study and consideration of different models of intelligence provision and technical cooperation, may be an important first step in helping to redress the information gap between industrialised countries and the Third World. Current practices for the provision of information through market mechanisms are far from perfect in responding to these problems. The creation of more collaborative structures among developing countries and new institutional capability would address these problems in part. A recognition of the intelligence function, above all, is crucial and the creation of an awareness of its importance for negotiation and planning purposes is an important initial step. Such efforts may in the near future be the most successful moves on the North-South chessboard.

## Notes

1 *Information and Economic Behaviour*, Stockholm: Federation of Swedish Industries, 1973, p. 11.
See for further clarification Stephen P. McGee, 'Information and the Multinational Corporation: An Appropriability Theory of Direct Foreign Investment', in J. N. Bhagwati, ed. *The New International Economic Order: The North-South Debate*, Cambridge, Mass. MIT Press, 1976.

2 '. . . the very large impact that computers are having on business and governmental organisations . . . is a profound change in the decision making process, which is being formalised, made explicit, and subjected to deliberate planning'. Herbert A. Simon, 'The Consequences of Computers for Centralisation and Decentralisation', in Dertousos and Moses, *The Computer in the Modern World*, Cambridge: MIT Press, 1979. See also Robert Gilpen, 'The Computer in World Affairs', in the same volume.

# 1

# The Political Economy of Information in a Changing International Economic Order*

Rita Cruise O'Brien and G. K. Helleiner

The central focus of economic activity in 'post-industrial' society is shifting from manufacturing of objects to the handling of information and knowledge.[1] The power of major transnational firms now rests as much upon their capacity to marshall information and knowledge as on their traditional role in directly productive activities. There is every indication that the sharpest aspect of competition in the future may be based more on the efficient use of specialised knowledge, information and new technological capacity for its communication and use than on more traditional factors. Information handling capacity already offers industrialised countries and firms considerable economic and political leverage in North-South interaction. Information vital to developing countries is frequently concentrated in the capitals of the North.

What is known or can be readily found out by firms and governments in industrialised countries about resources, weather, technology, market conditions, or indeed about information itself, far exceeds such knowledge in developing countries. Much of this information which is gathered and systematised for retrieval is not available outside its storage point. Universities and research institutions as well as privately financed research and development activities thus give public and private decision-makers in the North considerable advantage in their capacity to profit from world market participation and to bargain internationally. Negotiators and market participants from developing countries can thereby be placed at a serious relative disadvantage. Their informational difficulties derive not only from data scarcity and difficulty of access in one-off situations, but also from the absence of relevant expertise at either the national or intergovernmental level which can serve their interests on a continuing basis.

The expansion and reduction in the cost of facilities for communica-

* This chapter is a revised version of an article which originally appeared in *International Organisation*, Vol. 34, No. 4, Autumn 1980.

tion, transportation, financial transactions and travel have broadened the character and pace of global interaction and thereby altered the environment of politics. In particular, these developments blur the distinction between 'high' and 'low' politics, producing an even more complicated international 'chessboard' than was suggested by Stanley Hoffmann more than a decade ago.[2] The question of information access, which was formerly quite distinct from political calculation, has acquired major political importance in the new environment. But the literature on international politics and the relevant social science literature on information still examine the *effects* of increased communication and information flow as if they were simply accelerating the creation of global 'interdependence', without considering the implications of unequal informational access either for the efficiency and equity of international market functioning or for the creation of new forms of asymmetry in international bargaining capacity.[3] The emergence of 'North-South politics' and the growing pivotal importance of information in the world economy together add further credence to the frequent assertion that world politics have been changing too rapidly for conceptual paradigms to have kept pace.[4]

In this paper we can do no more than begin the process of analytical construction in the issue-area of specialised information and suggest conceptual refinements which might be made on the basis of systematic examination in various areas. In the next section we consider the role of information in the functioning of markets, employing the emerging literature in this area in economic theory; and follow that with the inequality of access to information in the international relations models of 'interdependence'. We then begin to develop a structure for analysis of the political economy of information as it relates to North-South interactions and relationships, noting some of the policy innovations of recent years and their limitations. The final section considers the implications for the developing countries and North-South relations of technological change in the information industry, the so-called 'information revolution'. There follows a brief conclusion, suggesting possible areas of research and policy analysis.

International economic and political systems have evolved in recent decades from a collection of closed hierarchical systems to a complex set of relationships involving a variety of new forms of interdependence. Information access among dependent countries was largely irrelevant in colonial systems and also largely so where there were dominant powers enjoying spheres of influence. But political independence and even partial delinking from the transnational economic system through nationalisation and other measures have radically changed the environment of information need and access. Countries of the South pursuing economic policies based on their own interests have new requirements, which have implied the search for new sources of information and expertise, new information systems, means of codification and use.

These countries and their major economic institutions have often entered forms of bi-lateral and multi-lateral bargaining in which their lack of information and expertise have added to their vulnerability in the search for establishing alternative markets and sources of supply. In no part of the current debate over global political and economic problems is the importance of information more dramatically evident than in the search for new forms of mutually agreeable relations – new contracts – between transnational corporations and developing countries.[5]

Nearly two decades ago, the study of information and knowledge seemed to occupy 'a slum dwelling in the town of economics'.[6] While there has been considerable progress in the literature of the economics of information since then, very little of it seems to have percolated through to the new interpretations required by international economic relations. Information is both a key intermediate input (which can be acquired at a cost) and an important factor of power in international bargaining. Differential access to information can therefore be a major element in the determination of the distribution of the world's income. Research on present stocks and flows, including new technological capacity for its collection, storage, transmission and use, can no longer be postponed.

## Information, markets and economic theory

Information and knowledge can take many different forms. Considerable attention has been devoted in recent years to the developing countries' dependence upon imported knowledge in the form of technology and to the imperfections of technology markets, the inappropriateness of much of what is available to them, and the need for developing increased indigenous technological capacity. Rather less attention has been devoted, however, to the more general question of 'market information'[7] or 'commercial intelligence', upon which the effective functioning of arms-length systems of exchange depends; this at a time when increased resort to arms-length exchange is an accepted part of Third World aspirations. This type of information cannot be entirely separated from knowledge of underlying technical change, demand shifts, and the like. The distinction between 'market information' and other types of knowledge is nevertheless a real one. A consideration of the economics of market information (and this is only one aspect of the information issue) must address such matters as the economics of alternative organisations, and the economics of search, as well as the peculiar properties of information markets. The remainder of this section is devoted to such a consideration.

Fully informed, rational economic man can be found in elementary economics textbooks where still 'his cool, consistent mind quickly and costlessly scans the myriads of alternatives facing him'.[8] In the popular mythology his principal habitat today is in large business organisations and, above all, in the transnational corporations. But, increasingly, economic theorists are investigating models of a more realistic world in which one typically finds instead 'the uncertain decision-maker acting under severely restricted conditions of information embedded within a communication system upon which he is becoming increasingly dependent'.[9]

Indeed, incomplete information, and resulting risk and uncertainty, have become central to major theories of economic organisation, both at the macro- and the micro-level. At the macro- or societal level, the debate between advocates of central planning and decentralised market systems has long centred on questions of data collection and distribution. Hayek, in the pre-computer age, believed that the sheer size of the data collection, communication and processing requirements for effective centralised decision-making made it infeasible whatever other merits or demerits it might have. Markets are themselves, he argued, the most efficient and effective information systems.[10] Needless to say, the new technologies of the electronic age today require the formulation of such arguments a little differently. The efficiency with which different macro-systems of economic organisation employ available data for the purposes of allocation of resources continues to be a matter for (pure) economic theorising. To our knowledge, however, these investigations have not as yet addressed the implications for 'social efficiency' (in which 'efficiency' is defined not in terms merely of Pareto optimality but in terms of income distributional objectives as well) which derive from *differential and asymmetric* access to information on the part of actors within the system.

At the micro-level, the theory of the firm itself and its international economic involvement is dependent on assumptions regarding its effectiveness in the generation, management and use of productive knowledge. If one traces the principal elements of power of major transnational corporations, key recognition must be awarded to the availability and quality of information. The structure of the modern transnational corporation is in fact information-dependent;[11] its 'appropriability' of information[12] being perhaps as important as its R & D capacity and the control it exercises over markets and technology. In a recent symposium on the very frontier of the 'economics of internal organisation' its organiser remarked on the 'striking' fact that all of the papers were 'critically concerned with information in some form or other' and could be regarded as explorations into 'a variety of informationally constrained resource allocation problems and institutional responses that characterise firms, organisations, and groups'.[13] The replacement of the market through integration or merger takes place, among other reasons, in consequence of 'transactional failures' in markets,[14] failures traceable primarily to problems in the processing of information. This is as true of the transnational corporation at the level of world markets as of the national firms for which the theory was first developed. Dunning refers to these failures in the transnational context as 'cognitive imperfections'.[15] The main strands of micro-economic interpretation[16] are based on two propositions in respect of the cost and quality of information processing.

Firstly, in circumstances of great complexity and uncertainty there may be limits to individuals' or firms' capacities to receive, store, retrieve and process information faultlessly, or to transmit information to one another effectively ('bounded rationality'); in such circumstances internalisation of what would otherwise be arms-length transactions or learning by doing may be the most effective means of organizing activity. Internalisation may carry the further advantage of some scale economies.

Secondly, where there may be reason to doubt the total veracity or completeness of information being supplied by one's (armslength) source, particularly where 'small numbers bargaining' rather than active competition (which is to some degree an automatic 'policeman') is found, it may be necessary to internalise one's information sources to prevent what has been termed 'opportunism' (i.e. 'an effort to realise individual gains through a lack of candor or honesty in transactions');[17] this may also reduce the incidence of 'information impactedness' stemming not so much from opportunism as from inherent difficulties of specifying contractual arrangements.

Both the cost of information and its quality evidently may be influenced by the institutional arrangements through which it is collected, communicated and processed. In general, 'A situation in which anonymous agents deal with an impersonal market is not conducive to efficiency with imperfect information'.[18]

Information (in the sense in which the word is here being used) can be considered as a product like any other which can be acquired at a cost – either by 'making' it oneself or by acquiring it from another. In the standard literature on the economics of search[19] it consists merely of information as to the range of available price alternatives in a real world in which because of transactions costs of various kinds, the 'law of one price' is all too rarely encountered. Assumptions as to the distribution of prices and the costs of searching lead to decision rules both for buyers and sellers as to how much to search or how much information to offer through advertising. Matters become more complex when the searcher does not even know the degree of price dispersion, for then he cannot know how much to search; the ignorant, e.g. the tourist in a foreign land, can expect to pay more when he buys and receive less when he sells. But market information involves much more than this static and unidimensional fact of price dispersion. Prices and their degree of dispersion are constantly changing in response to innumerable influences upon demand and supply; market information therefore becomes obsolete very quickly. More important still, information as to the likely (though obviously still uncertain) future state of prices becomes a necessary input to market decision-making; even better, if one possesses the capacity to process it independently, is information concerning the major underlying influences upon future market behaviour, for this permits one to make one's own judgements independently of others' possible biases or 'opportunistic' behaviour. Information of this more qualitative kind is clearly much more difficult to quantify or even theorise about. Of at least equal importance to price information, particularly when goods are not homogeneous, is information as to the *quality* of items to be purchased. Search may therefore take place on the intensive (quality) margin as well as the extensive one. In some instances their suitability to the buyer can only be tested through use; such products have been termed 'experience' (as opposed to 'search') goods.[20]

Among the means of reducing qualitative informational uncertainties in a complex world are the development of 'customer relationships', brand loyalties, 'reputation' and 'goodwill'. In conditions of great information uncertainty, cautious buying (or selling) behaviour is understandable: the familiar – in terms of existing relationships, geography, language, etc. – may be chosen even when the cost, objectively speaking, appears high. As information improves, the breadth of shopping (and marketing) increases. Thus what geographers call 'information space',[21] the nature of information networks, can play an important role in the determination of the direction of both domestic and international flows of goods and services.

Those who do not themselves possess the information they require must acquire it from others. While such acquisition does not always occur on markets, it is analytically helpful (as in the case of technology) to consider it as if it did. Information as a product and information markets, like technology and technology markets, have peculiar properties.

1 Information is not an exhaustible product; that is it is not 'used up' through employment or expanded dissemination. Many can possess it at the same time. Its provision to others, even if transactions costs are zero, may nevertheless involve one in losses, and there may be advantages in not sharing it. On the other hand, substantial costs may also be sensibly incurred for the sake of gains which derive from the provision of information to others – through advertising. (Indeed, as has been seen, there may be advantages in passing on *mis*information; and that fact necessitates a greater capacity to assess information acquired from 'external' sources.)
2 The cost of transmitting information which has already been produced (transactions cost) is not zero, and can be expected to vary

greatly as between different types of information and different means of transmission. Transmission systems have their own factor-intensity and scale requirements, and it is likely that the lowest-cost systems require substantial capital inputs and scale. The latter create barriers to entry to certain transmission systems, and it follows that different actors will have differential access to low-cost media.

3 The returns from the production of information are generally not fully appropriable; that is, it is often possible for others to realise gains from the use of information which you 'produced' without your being able to obtain a share of them. There are no laws of conventions, such as those governing patents and trade-marks, which protect intellectual property rights in this area. It follows that great quantities of information are transferred through low cost or even totally non-commercialised channels, sometimes quite informal ones. (Some kinds of information are nevertheless not easily transferred, such as those learned best by doing; and others are successfully kept secret.)

4 Information may be a product of extremely rapid obsolescence, an attribute which also impedes its effective transfer among unrelated decision-makers.

5 Like technology, its quality is impossible to judge accurately until one possesses it; many of those with the 'best' supplies are themselves, consciously or unconsciously, biased with respect to its content. 'Opportunism' and conflict of interest must make one particularly careful about its reliability.

6 Information is frequently only available (at reasonable cost) in packages or indivisible lumps which, not being tailor-made, include goods and services other than those actually sought; these indivisibilities may relate to the size of the information package itself, to the packaging of the required information with goods or other purchases, to the minimum period of time over which it is to be offered, etc. It is often a by-product of some other activity.

7 On the supply side, the production of information seems likely to be characterised by economies of scale, economies of experience (learning by doing), and positive externalities; the cumulative effect of acquiring information over wider areas and over longer periods of time is to render one relatively better and better at acquiring more. 'Since the cost of collection of information is (approximately) independent of its use (although the cost of dissemination is not), there is a strong tendency towards monopoly in the provision of information: in general, there will be a 'standard' source for trade information.[22] Still on the supply side, the production, storage and processing of information is highly skill-intensive and capital intensive.

8 The longstanding legal precept, *caveat emptor*, places the responsibility for the acquisition of information firmly upon the buyer. Salesmen are not obligated to provide balanced assessments of the

merits and demerits of what they offer. Once the sale has been concluded, moreover, there is no provision for renegotiation of terms on the basis of information subsequently acquired by the buyer; the sanctity of contract now takes over. (Nor are there markets on which one can purchase insurance against the possibility of having made a mistake; in larger organisations the law of large numbers offers some such protection but for smaller ones, the implications of mistakes can be devastating.) In some areas, sellers attempt some self-discipline through the formation of 'better business bureaux', 'chambers of commerce' and the like; and in others the consumer movement has made some in-roads on both laws and practices with respect to the prior provision of information, guarantees, and the improvement of original contracts.

## Information, power and interdependence

In research and writing on international relations of the 1970s, two major trends emerged. Trying to improve on the old 'realist' school which dominated models in the 1950s and 1960s, American scholars in particular have developed models of complex interdependence, giving new prominence to interaction based on 'low politics' rather than pre-eminent attention to military security and force.[23] From a different perspective, Third World scholars, in particular those from Latin America, began to examine the constraints on development posed by the economic dependence of peripheral economies.[24] Interdependence theorists have been more concerned to describe and measure rather than explain the consequences of interdependence, working mainly on the relations among industrialised states. It has therefore been somewhat difficult to interpret the costs of interdependence for weaker states in this approach, which is, of course, the main concern of dependence analysis.

In focusing on the terms of unequal exchange between rich and poor countries, the dependence approach gives more weight to nationalism – not as a felt cultural need or a valued end in itself – but as a *realpolitik* term employed to try to seek change in the present organisation of global capitalism. The search for amendments to the interdependence approach has in most recent formulations sought to reintroduce the element of power,[25] and particularly the power of the nation-state in relation to transnational economic forces and transnational firms. Despite its shortcomings, the old realist school directed attention to states as significant actors in international politics: its narrow approach to power based on military strength may now be applied to the consideration of hegemony based on resources, industrial capacity, technology and the control over markets.

The proposal for change in the international economic system by the developing countries is founded on a broad-based approach. Despite their manifest weakness and vulnerability in certain forms of bargaining, these countries have largely accepted the inevitability of interdependence. Changes in the world economy have never evolved in consequence of competition among equals, but rather through the emergence and changing influences of great national economies, some of which have become predominant. A recent historical survey of interdependence charts its 'ebbs and flows' in something of a cyclical pattern,[26] rather than its existence as an ever growing phenomenon, despite technical and economic changes in the global system, political pronouncements or scholarly attention.

In consideration of information inequalities and their effect on power in international bargaining, certain trends have become obvious which require a reconsideration of the terms of present analytical constructs of interdependence. The aim of developing countries to assert a new form of political action over transnational economic forces and to demand a more holistic approach to the formulation of broader policy bundles has made decision-making increasingly inter-sectoral and put incremental strategies at a disadvantage.[27] There is, therefore, a notable gap between the search for new international planning measures with the aim of restructuring international relations, and the achievements in specific sectoral negotiations wherein the technical and specialised parameters are left to qualified negotiators.[28] It is precisely at this level where achievements of the South in the 1970s have been minimised, in part because of lack of access to information or poor organisation and use of information. A crucial question for developing countries is how power over resources may be converted into power over events: in many instances actors may not be able to make effective use of the resources under their control in order to bargain effectively. This has serious consequences not only because of the high cost of establishing alternative markets and sources of supply, but because 'increased interdependence between unequal partners may lead to a greater loss of domestic policy-making capacity among the weak'.[29]

Interdependence theorists have tended to concentrate on the consequences of such relationships, questioning whether disparities in economic capacity or 'vulnerability' necessarily lead to inequitable bargaining outcomes or a permanent international hierarchy. Realists and dependence theorists claim that the outcome can always be predicted from the relative power position of national actors. None of these approaches is entirely satisfactory. The concept of 'asymmetric interdependence' allows for reciprocal interaction which is not based exclusively on mutual interest but nevertheless results in mutually acceptable resolution of conflict. 'Less dependent actors', we are told 'can often use the interdependent relationship as a source of power in bargaining over an issue and perhaps to affect other issues'.[30] This can be formulated as a simpler proposition: that the *more powerful* can use the interdependent relationship as a further source of power. New forms of interdependence (whether asymmetric or not) in areas of 'low politics' may be seen as justifying the present systems of interaction in which the 'rules of the game' or the parameters of negotiation are effectively determined by the more powerful actors.[31]

The new agenda item of information in world politics suggests the necessity of analytical distinction according to the *capacity* of the actors in an interdependent situation. That is, the inequality of access to information adds a new dimension to asymmetry which may be originally based on political and economic divergence – it is a divergence in the capacity of the parties involved. Having better access to and control of information in a negotiating situation is an important factor of power; having poor access to information tends to incur present and future costs for the more vulnerable parties. Some specific examples will help to clarify the problem.

The manpower and expertise which can be marshalled by industrialised countries in preparation for international negotiations such as those of the GATT, the Law of the Sea, or the WARC[32] far outstrip those which are available to developing countries.[33] This international form of inequality may be detrimental to the global interest, not just to that of the developing countries. There is growing concern that the lack of information to permit adequate preparation on the part of developing countries for intergovernmental negotiations makes the bargaining process unworkable. Certainly it leads to the politicisation of issues in the absence of adequate technical analysis. And because of the resulting incapacity to sustain and specify negotiating positions in meetings which take place at certain intervals, it may jeopardise results previously achieved.[34]

Discussions on the reform of the international monetary system, or on the allocation of resources of the sea and space are thus often pre-empted by the more powerful nations, leaving the developing countries with only a 'yes, but . . .' role or encouraging them to politicise issues as a substitute for the technical arguments which they did not possess.[35] Such acts have been regarded as 'obstructionist' by the more powerful actors in international bargaining but may, with the hindsight of nearly a decade of new forms of North-South bargaining, be seen instead as having been quite astute. Might they now not be considered as (variably effective) delaying tactics which gave much needed time for the preparation of more well informed and effective positions?

Interdependence may be accepted as inevitable by developing countries, but the specific terms on which it is to be achieved are still at issue. Those who argue that it is an inexorable trend, based on a technocratic justification of the future system, may be regarded with scepticism by the South:

> Sociospheres, econospheres and technospheres are the complex and interwoven network of institutions, organisations and inter-dependent technological systems which begin to show potential as a unified network of human service systems around the planet – depending for its integrated functions on a swift availability of and access to, shared information and communication.[36]

Its espousal by the politicians of the rich countries who argue that it will benefit all and resort to symbolism which suggests that conflicts of interest are *passé*[37] may be regarded with mistrust. But, above all, its promotion as a functional approach[38] to resolving the problems of a new information era in the interests of the already information-rich states is likely to be resisted.

The politicisation of the information issue is already apparent, and a growing source of concern to US policymakers:[39] 'Trade doesn't follow the flag anymore', according to a member of the US Mission to OECD, 'it follows the communication system'.[40] Global information issues may be schematised in terms of 'protectionism' versus the 'free flow' of information. As in the analogous case of international goods trade, national interests and the interests of transnational business line up differently on this issue depending on whether information is in a raw or processed state.

Transborder data flow has become a major new information issue: data on markets, technology and credit assessments, all of which are vital to the success of private firms carry a potentially high market value. It is not only a probable source of monopoly power, but a further challenge to sovereignty. Given the US near monopoly of data bases and equipment, transborder data flow, even in a raw variety has become of major concern to European countries. Study of these questions continues within the OECD, and was raised in the context of the general debate on services at the meeting of the General Agreements on Tariffs and Trade in late 1982. It was placed on the agenda by the United States in an effort to maintain the free flow of data traffic, which is in the interests of its major transnational firms in services. Some newly industrialising countries have become aware of the importance of data flows, and have introduced restrictions on data traffic and developed national policies on transborder data trade. Brazil has the most comprehensive legislation, and Mexico has rapidly followed this approach. Algeria also allows no transmission of computer-held data to other countries.[41]

International business is quite concerned about these developments, as, in their view, such moves seriously restrict the environment in which transnational firms seek to work. It may, in their view, restrict private foreign investment in countries with restrictive legislation. Transborder data flow is beginning to emerge as an important North-South issue. Its watching brief is held by the Intergovernmental Bureau of Informatics

(Rome) which has become a vehicle for the consideration of the interests of the South in these matters, as well as a forum for the consideration of the use of information systems in development.

International telecommunication policy has provided another arena of international debate which has begun to concern the countries of the South. The electromagnetic spectrum and the geo-stationary orbit (at 22,000 miles above the earth) are the resources used by broadcast, telephone and satellite communication. The first reallocation of the spectrum for twenty years was achieved in 1979 by the World Administrative Radio Conference of the International Telecommunication Union. The ITU and its specialised committees which have functioned in the past to guide and manage international telecommunications are increasingly regarded as inadequate by some recent members among the developing countries. Previous policy on tariffs and channel allocation, vital to international business, has been established easily because the spectrum was not yet regarded as a finite resource. Frequency allocations in the past have been made on a first-come, first-served basis.

The spectrum is now recognised as a limited natural resource, which had previously not been challenged in terms of access for commercial, state or military needs. In 1979, developing countries laid claim to as much of the non-broadcast spectrum as they can by demanding 'parking slots' which they could not yet use. The development of better bargaining capacity as well as resource allocation to national planning in this area is crucial to the protection of developing countries' interests.

The limited capacity of developing countries effectively to put their case in the face of much greater in-house expertise among business and government representatives from the industralised nations, as well as the structure of negotiation itself, are still major barriers to the kind of changes which might win long-run acceptance. The deliberations of the WARC postponed a number of important matters, leaving certain fundamental issues for discussion in future ITU meetings, by which time developing countries may have further opportunity to assess their requirements and achieve better briefing about technical details. In the past, their incapacity to assess future needs has put a particularly dramatic twist on such negotiations, wherein future disadvantage may have been negotiated without adequate preparation.

## Information, developing countries and North-South relations

Recent attempts by individual countries of the South to alter international economic relationships through nationalisation, the exercise of greater control over exports and imports, unpackaging, etc., can be interpreted as altering, in a fundamental way, the required organisation of the relevant information flows and processing systems. It becomes necessary to assess the capacity of the informationally weak to acquire, usually at arms length, and to process the information which they now require in order to make independent decisions which are in their *own* interest, rather than in the interests of the total information (and other) systems of which they were previously part. Difficult decisions must now be taken with respect to how much to spend, and in what ways, in order to improve the 'efficiency' of *national* decision-making and the effectiveness of national participation in international bargaining.

In analytical terms it is important to distinguish various types of interaction (or negotiation) in each of which there are different informational requirements:

- private firm to government (e.g. bank borrowing, transnational corporate relationships, commercial transactions),
- private firm to private firm (e.g. daily commercial transactions),
- multilateral government to government (e.g. GATT, WARC),
- bilateral government to government (e.g. international commodity agreement negotiations).

It has been usual for economists to specialise in the analysis of the former two types and for political scientists to specialise in the latter two. Our own approach has proceeded along this traditional path. But we have come to see that the same informational issues – political and economic – arise in them all.

Country capacity is also central to the analysis of specific informational issues and developing country needs, as they relate to North-South interactions and negotiations. Differentiation must therefore be made among:

- those countries which have already built or readily can build up a national negotiation capacity, at least in specific sectors, (e.g. Brazil on finance),
- those which must generally rely on outside expertise because of an indigenous expertise constraint, (e.g. Tanzania).

Country capacity may be further differentiated as among resource-rich, expertise-poor countries, where wealth can be substituted for manpower resources and expertise purchased (the oil surplus Arab States); and resource-poor, manpower (relatively)-rich countries where indigenous capacities of expertise can be marshalled to bargain successfully in crucial sectors (Jamaica on bauxite, for example).

It is also important to identify those cases where differential country capacity can be overcome to the joint gain of all by the sharing of information, e.g. OPEC, CIPEC, the Bogota Group. The economic characteristics of information – notably non-exhaustibility, scale economies, experience economics and positive externalities – imply substantial benefits from joint development of information facilities where the specific needs are similar.

The consequences of imperfect information are nowhere more dramatic than in the resource sector. Governments of developing countries are asked to contract with foreign transnational corporations for many years in advance with respect to the terms of resource exploration, exploitation and development. At the time the bargains are originally struck it has been usual for the private firm, however uncertain it may be with respect to the final outcome of its explorations, to be substantially better informed as to the probabilities (and the terms of similar agreements elsewhere) than is the host government. As exploration proceeds, and especially if development is undertaken, however, the information available to the host government increases and the informational imbalance declines. Private firms understandably seek to protect themselves against their risks and uncertainties, including those associated with future government policy changes, by demanding high rates of return, at the outset. Governments equally understandably, seek to utilise their new information, as it arrives, in order to improve the terms of the bargain which they originally made when they were relatively badly informed.[42] The impasse which has developed between host governments in developing countries and the resources transnationals is said to have produced a marked decline in resource exploration and development in the Third World in recent years which is in the interests of neither. The remedy would seem to lie in improved (and presumably socially subsidised) means of providing the relevant information – particularly to the host country governments which originally most lack it. To some degree, these host countries have already managed to purchase much more of their requirements from specialised firms than they used to.

Access to private international lending was vastly enlarged in the 1970s for many developing countries; yet little is known as to how the sixty or so countries who are new to the market learn about and keep abreast of events and experiences in the rapidly changing world of international finance.[43] It is possible for countries without sufficient information or knowledge to make very costly errors, because of complexity and variability of conditions in international financial markets (variation in exchange rates, interest rates and management fees, to mention but a few issues).[44] There is no single institution which can provide continuous specialised information to new developing country borrowers, which might be analogous to the information sharing advantage of established bodies like the Bank for International Settlements, serving the central banks of the industrialised countries.

The provision of information has become a lucrative business for large investment and merchant banks, the services they provide ostensibly kept entirely separate from their interest in lending. Among the best known advisory services are those provided by the 'triad' (Kuhn

Loeb Lehman Bros International, New York, Lazard Frères and Co., Paris, and S. G. Warburg, London), although many other banks have recently set up similar services.[45] The question for developing countries is whether it would not be in their interests to establish alternative means of expertise and advice, form their own information consortium and enable better control over advisers.

Buying or selling on world markets requires that developing countries try to optimise their means of search in the manner of an arms length shopper. Too few studies have considered new and alternative forms of marketing commodities or manufactured goods: marketing literature remains largely unaffected by the many recent changes in world trade. Little is known, for example, about how newly industrialised countries (South Korea, Singapore) acquire their information concerning global markets. Many of the poorer countries have become involved in forms of market management over which they have lost significant control because of their lack of information on pricing for example. Information is required in order to know where to acquire information and knowledge of specialised markets is an essential asset to better bargaining. Where intrafirm trade still persists in international transactions, the search for developing countries may be conducted instead for the purpose of monitoring and controlling transfer pricing.[46] At the present such information flows to the developing countries' decision-makers through a wide variety of channels and media. It arrives through 'in-house' collection and communication on the part of overseas commercial representatives, attachés, or agents; it is assembled through trade journals, specialised consultancies, the services of brokers and dealers, the messages of salesmen and foreign trade bureaucrats, and various informal contacts.[47] Rarely is its assembly and dissemination systematised on an ongoing basis as it typically is in large commercial enterprises.[48] The basic informational resources of libraries, data banks, and the like available to the poorest countries are hopelessly inadequate: frequently taking the exclusive form of published sources arriving sporadically by sea mail to understaffed institutions.[49] For those 'in the trade', published sources, even 'hot from the press', are typically too obsolete to be useful for daily decision-making. The linking of computerised data banks and telecommunication facilities have altered the business information environment dramatically. While the present means of electronic data processing has in the short-run undoubtedly enhanced the efficiency of information flows among industrialised countries and their major business firms, still greater current and future problems arise for developing countries in deciding how to make information effective in its application. Their greatest disadvantage is in respect of the capacity to mobilise information and make it available in *usable* form at vital points in the process of negotiation or decision-making.

It is the application of knowledge and organisation, skill and planning in the use of information, whether produced or transmitted by ordinary or electronic means, which may be the most crucial factor in negotiating advantage: the capacity to synthesise information is itself based on innovation and is vital to negotiating power. This capacity to organise information in terms which are appropriate in respect of speed and specificity may be most effectively achieved in the short-run by buying-in outside expertise. At present this is available to developing countries seeking to improve their negotiating position either through UN agencies or different forms of bilateral aid, or through purchases on the market – consulting firms in particular. Acknowledged 'in the trade', there is as yet no systematic consideration of the advantages and limitations of the use of outside expertise in negotiation.[50] Host countries and firms are able to purchase specialised information from consultancy firms like Commodities Research Unit Ltd, or Chase Econometrics. While ensuring a greater flow of information, there is considerable difference in the capacity to use the information effectively among say copper transnationals like Anaconda, Kennecot (or their parent companies) and producing countries with limited expertise like Zambia.

Some new institutional initiatives in the spirit of the altered global 'objective function' described above attempt to enlarge developing country negotiating capacity. Specialised courses and conferences have been developed by the Centre for Transnational Corporations, UNCTAD, the International Legal Centre, the Commonwealth Secretariat, and UNIDO. There also exists a series of data banks designed to reduce the information gap of developing countries. Until recently they have been most useful for project design where there is no time constraint, or for the provision of background information for research purposes in relation to national decision-making and an improvement of aid requests. Recently there has been increased provision of international financial and trade data, through a system funded by the UN.

DEVSIS (International Development Research Centre, Canada) is designed primarily to assist the design of development projects. UNISIST (UNESCO) helps developing countries establish their own information systems for use in national and local planning. The UNIDO system provides retrieval of information on sectoral and plant level industrial design. The Development Information Network (UNDP, New York) provides current financial and trade information, but does not have a 'memory' for the consideration of long-term trends. The problems of these systems are chiefly their under-utilisation which is in part explained by their attempt to achieve universality. In contrast, systems built on a coherent international professional group like the data banks of INIS (nuclear sciences) or AGRIS (agriculture) serve specific research needs and are much more effectively used.

There is, of course, a danger in the field of storage that the UN system and others seek to create data banks without specific reference to professional capacity for decoding, synthesis and use of information for sectoral requirements. These may generate inordinate expense on large prestige systems hardly used at all where the needs are greatest. For the purposes of negotiation (or other uses) it may be essential that the information be location-specific. There is a considerable differentiation in efficiency among user-specific information and that which can be effectively derived from data banks.[51] Thus, while attempting to be as relevant as possible, most of the current data banks inevitably fall short of the specificity demanded of most business and other negotiations.

## Technological changes in information and transmission systems

The most important effects of the current technological changes in the information sector are:

- the cost reductions achievable through miniaturisation based on microprocessors;
- the increased volume and specificity of improved computer systems;
- the enhanced speed of global transmission through new telecommunication systems based on optical fibres and the use of satellites.

The latest high growth area in the industry is the convergence of electronic information processing, data-banks and telecommunications networks into a single system called telematics. There is also the rapid expansion of digital data networks and the use of video channels. Expansion of use in industrialised countries is partly assured because of relatively low energy consumption compared with earlier forms of information and communication technology. The basic question which must be raised in relation to technological change in the information industry and its potential effect on global information access is whether these new developments will (i) *increase* the already substantial potential for centralisation and control of information in traditional world centres thus enchancing their power and leverage, or (ii) be susceptible to marshalling by developing countries (individually or in regional or sectoral groups) in order to *decrease* their present information disadvantage?

An adequate response must be related, most fundamentally, to the time horizon to be considered and the capacity of systems to relate available information to specific requirements. In the short-run, technological developments may increase the power of already data-rich countries and firms. For the present, commercial incentives dictate the design of new equipment and systems and these are therefore geared to large markets: private firms and carriers and governments in industrialised countries. Transnational corporations are dependent on multinational computer systems to increase their manoeuvrability in their relationships with governments, suppliers and customers. It is acknowledged in the business literature[52] that the wider use of these systems will enhance their power and influence in relation to governments and will widen the gap between rich and poor.

Far ahead even of Western governments in their utilisation of the most modern information systems, major transnational corporations have long since overcome the major imperfections of market exchange by internalising their informational (and other) flows. These firms not only have the capacity to utilise the new hardware of the data processing industry but also to develop or to encourage others to develop the even more important, and often industry – or firm-specific 'software' – the 'automated filtering and screening mechanisms' for information relevant to them.[53] Market incentives are, as always, guiding further technological developments in information processing:

> the large multinational firms with widely scattered operations represent the most attractive customers (for computer firms), so that enormous investments are being made to develop new applications and capabilities that would appeal to these firms. Further, the computer industry itself is becoming more unified worldwide through the use of common procedures, systems and software.[54]

Modern telecommunications systems have thus evolved under US leadership, and specifically in response to the needs of transnational corporations (and defence surveillance). Transnational enterprise depends on a system of 'free flow' of information and open borders. The reduction of non-technical barriers like tariffs on 'alternative voice-data circuits' has led to the immense growth of private carriers since the end of the 1960s upon which many forms of transnational enterprise (and business consortia) are dependent.[55] These tendencies have led certain analysts to ask whether the international information policy-making process has been able to deal adequately with technological breakthroughs which promote increased interdependence.[56] Few have asked the basic question – in whose interests and at what cost?

In the longer-run, technological change and its effect of cost reduction may be able to reduce the data dependency of developing countries and facilitate the efficiency of marshalling information resources in relation to their interests. Much depends on its specific application to sectors where needs can be identified or collaborative effort explored. There is a growing literature on the uses of information technology and development. Among the best contributions policy studies done by the Information Policy Group in UNESCO which directs the UNISIST information system. Its approach is based soundly on national capacity and on the links between electronic and other information systems at national level.

Questions about production and transmission must be related to specific information needs, and its most effective use. As in the case of information generally, the greatest problem in the improvement of information access through electronic means may be in the knowledge and institutional infrastructure required to make use of the information system. With the substantial reduction of hardware cost, it is possible that expertise of electronically produced software, whether bought in or developed locally, will henceforth be the dominant cost. Are there enough incentives for it to be made commercially in forms which will be useful to meet the specific needs of developing countries? Can it be suitably adapted for these purposes from other systems? Can it be obtained from non-commercial sources?

A powerful impetus to the rapid accumulation of information systems in developing countries may be seen to come from the moves to promote self sufficiency in computer technology[57] and to add informatics[58] to the agenda for international discussion. The potential danger of such general initiatives is the mistaken assumptions drawn from other economic sectors that telematics systems may in themselves promote growth and development, and that the only problem is their rapid acquisition. The use of high technology in other sectors has frequently produced new forms of dependence rather than enlarging independent capacity. It has even generated new dimensions of inequality and poverty, when balanced against the opportunity cost of investing in other sectors. Developing countries cannot ignore changes in information production, but will they enter this field with the care that experience in other sectors would suggest is required?

Informatics cannot in itself reverse the traditional concentration of knowledge and expertise in the capitals of the North and major transnational firms. As the cost of hardware production is progressively reduced, there is already evidence for concern that relatively low-cost information systems will be 'unloaded' on developing countries in return for lucrative software contracts. Already the returns from expenditure on information systems in advanced industrialised countries are under critical examination. Among the key questions raised is the adequacy of planning and professional capacities in relation to the electronic processing and information storage capacities. Sometimes the identified limitations in this respect are only 'teething problems'; sometimes, however, they have proven to be much more fundamental ones involving large financial loss or even a weakening of decision-making capacity. Such problems are both more likely and more costly for countries of limited income.

Informatics may not at all aid the integration of technological change at national or international levels. It may create a new form of dualism in developing economies generated by professional engineers and systems analysts who seek to define their own professional advancement in terms of new gadgetry and prestige expenditure. There is also a risk of producing, codifying, and storing false or dubious information if the quality of national statistics remains as poor as it still is in many countries. 'Garbage in' is still 'garbage out' in the most sophisticated of information systems. There is even a serious risk of creating unnecessary unemployment by off-loading inappropriate electronic information processing systems in service sectors like banking or in government, where accountants, clerks and messengers can function quite satisfactorily. It is not at all apparent that the new forms of information will be any more reliable.

In short, it is unlikely that in the near future the new information technologies will significantly expand development capacity in their application domestically. Astute choice from what is available on the market and above all a better design of software specifically in relation to local sectoral needs can lower the risk of a serious waste of resources in this sphere. It may, in the short-run, be best simply to build up knowledge of the relevant expertise market which can provide information rather than to spend limited resources on independent user-specific software systems. There is a host of intergovernmental issues which are raised by recent developments in the technology of information production and transmission. Some awareness of these issues has been created among developing countries by the Intergovernmental Bureau of Informatics (IBI) has provided a venue for the discussion of some of these issues.[59]

In specific sectors like mineral negotiations in which there is great common interest, the political initiative essential to information exchange across national boundaries still seems distant. For the immediate future the bargaining capacity of developing countries in respect of these issues is seriously limited by such problems. At both national and intergovernmental level an enlargement of information handling and processing capacity may turn this situation around. But the present enthusiasm for informatics claims too much both in terms of national planning and intergovernmental capacity. Information systems will not themselves promote growth and development; nor will they necessarily strengthen the South in bargaining terms. Serious consideration is required concerning where and how computer and telecommunication resources can be best deployed. In the meantime caution and critical reflection may still forestall the creation of new and unnecessary forms of dependence and disadvantage.

## Conclusion

In this *tour d'horizon* of the information issues in North-South relationships we have raised significant questions for discussion, to be taken up by specialists in further chapters. Our primary purpose in writing the original article was to call attention to the importance of these issues, and to stimulate further thought and research on these questions. We are particularly interested in the case for a more integrated approach to the relatively neglected informational issues in the ongoing discussions of new international economic orders, one which draws both on the already quite rich literature on the economics of information and on the changing models used for interpretation of international relations by political scientists. It seems to us that in this sphere, as in so many others, full advantage of the analytical and research opportunities offered by existing approaches is not being realised. A more thorough political economic analysis of the role of information raises certain theoretical problems for reconsideration, among them 'market imperfections' and 'asymmetric interdependence'.

A most immediate task is to consider how precisely the information factor impinges on a variety of international commercial relationships, in which product cost and accessibility are integrally linked. Studies of the information industry itself, its structure, product cycle and the features of technological change, will enlarge our understanding of this core sector.

In the new international division of labour, it is assumed that high technology industries, especially those contained in the information economy will be predominant in industrialised countries. And yet, those sectors of the world economy producing raw materials or manufactured goods remain dependent on the means of information and telecommunication to enhance their range and capacity for commercial exchange. In our original article we called for sectoral studies of information access and international negotiation; we called for empirical categorisation on stocks and flows in these sectors. The following chapters of this book make an initial attempt to do what we were seeking in that article. We still believe that an understanding of the information environment in different forms must precede an analysis of technological changes in information and their effect on capacity.

Problems of immediate policy relevance to developing countries may be rectified by thought and action with respect to the provision of low cost usable systems of market information, particularly for the benefit of the poorest and weakest countries. By no means all information is sector-specific or ministry-specific; yet it is frequently collected in parallel efforts through different sources in a wasteful and repetitious manner, without adequate domestic distribution or use. Information is often acquired at great cost from foreign sources when it is already readily available at home (or in cheaper foreign sources known to other domestic users). All of these general suggestions for international and national reforms are analogous to those made in the realm of technology.

Continuing discussions on the prospects of creating an intergovernmental secretariat, like the OECD, for developing countries[60] increase the relevance of considering the prospects of strengthening information-sharing and expertise among developing countries. There may be ways to take advantage of externalities and scale economies and the exchange of information so as to outweigh those possessed at present by large, experienced commercial firms. The stimulation of competition among sellers and cooperation among buyers will improve the information flow. In addition to improving the prospects of sharing information, a rationalisation of the current expertise market may be achieved. With improved conceptualisation of information issues in political economic terms and detailed research at the sectoral and country level, it will be possible to analyse the enlarging range of information policy issues which lie before developing countries.

## Notes

1 Knowledge has therefore become a major factor of production. 'As a consequence, the "knowledge industries" – education, information processing and research and development – have become the most important sectors of advanced industrial societies'. Robert Gilpin, *US Power and the Multinational Corporation*, New York: Basic Books, 1975, p. 166.
See also Marc Uri Porat, 'Global Implication of the Information Society', *Journal of Communication*, Winter 1978, p. 73. Porat's more detailed study (9 vols), *The Information Economy: Definition and Measurement*, US Dept of Commerce, Office of Telecommunications, May 1977, estimated that in 1967 46% of US GNP was in information related activities, accounted for more than 40% of the work force (1970) and earned 53% of all labour income (1970).
The concept 'post-industrial' society was made popular, of course, by Daniel Bell, *The Coming of Post-Industrial Society*, London: Heinemann, 1974, the argument of which is based on the growing importance of the tertiary sector. A recent critical review of Bell, Dahrendorf and other 'modern improvement theorists' may be found in J. Gershuny, *After Industrial Society? The Emerging Self-Service Economy*, London, Macmillan, 1978.

2 'International Organisation and the International System', *International Organisation*, Summer 1970 (vol. 24, No. 3).

3 See for example R. Keohane and J. S. Nye, *Power and Interdependence*, New York, Little Brown 1977, John McHale, *The Changing Information Environment*, London, Paul Elek, 1976 and A. G. Oettinger *et al. High and Low Politics of Information Resources for the 80s*, New York, Ballinger, 1977. All note information inequalities without incorporating them into their models.

4 Conclusion in Robert O. Keohane and Joseph S. Nye, Jr, eds. *Transnational Relations and World Politics*, Boston, Harvard University Press, 1972.

5 'A key consideration in bargaining power rests with the extent of informa-

tion available so as to detect and evaluate the practices of transnational enterprises', C. V. Vaitsos, 'Money as a Negotiable Input in International Business Activities', unpublished (UN Centre for Transnational Corporations, NY), 1978, p. 31.

6 George Stigler, 'The Economics of Information', *Journal of Political Economy*, Vol. LXIX, No. 3, June 1961, p. 171.

7 G. K. Helleiner, *World Market Imperfections and the Developing Countries*, Overseas Development Council, Occasional Paper No. 11, May 1978, pp. 11–12.

8 M. Shubik, 'Information, Rationality and Free Choice in a Future Democratic Society', *Daedalus*, Vol. 96. Reprinted in D. M. Lamberton (ed.), *Economics of Information and Knowledge* (Penguin, 1971), p. 358.

9 Ibid. p. 361.

10 F. A. Hayek, 'The Use of Knowledge in Society', *American Economic Review*, Vol. 35, No. 4, September 1945.

11 Productive knowledge has become perhaps a more important element than capital in the operations of such firms. C. V. Vaitsos, 'Power, Knowledge and Development Policy: Relations Between Transnational Enterprises and Developing Countries', in G. K. Helleiner, *A World Divided*, Cambridge University Press, 1976, p. 140.

12 Stephen P. Magee, 'Information and the Multinational Corporation: An Appropriability Theory of Direct Foreign Investment', in J. N. Bhagwati, *The New International Economic Order: the North-South Debate*, MIT Press, 1976, pp. 318–319.

13 Michael Spence, 'The Economics of Internal Organisation: an Introduction', *The Bell Journal of Economics*, Vol. 6, No. 1, Spring 1975, p. 164.

14 Oliver E. Williamson, 'The Vertical Integration of Production: Market Failure Considerations', *American Economic Review*, May 1971.

15 John H. Dunning, 'Trade, Location of Economic Activity and the MNE: A Search for an Eclectic Approach', in B. Ohlin *et al.* (ed.), *The International Location of Economic Activity*, Macmillan, 1977, p. 403.

16 See also J. Hirschleifer, 'Where are We in the Theory of Information?', *American Economic Review*, Vol. 63, No. 2, May 1973; H. Malmgren, 'Information Expectations and the Theory of the Firm', *Quarterly Journal of Economics*, Vol. 75, August 1961; and J. Marschak, 'Economics of Inquiry, Communicating, Deciding', *American Economic Review*, Vol. 58, No. 2, 1968.

17 Oliver Williamson *et al.*, 'Understanding the Employment Relation: The Analysis of Idiosyncratic Exchange', *The Bell Journal of Economics*, Vol. 6, No. 1, Spring 1975, pp. 158–9. Williamson develops these ideas more extensively in his *Markets and Hierarchies, Analysis and Antitrust Implications, A Study in the Economics of Internal Organization*, The Free Press: New York, 1975.

18 Michael Spence, op. cit., p. 171.

19 George Stigler, 'The Economics of Information', *Journal of Political Economy*, LXIX, 3 (June 1961).

20 P. Nelson, 'Information and Consumer Behaviour', *Journal of Political Economy*, Vol. 78, March–April 1970.

21 Gunnar Törnquist, 'Comment', in B. Ohlin *et al.* (ed.), *The International Allocation of Economics Activity*, Macmillan, 1977, pp. 156–7.

22 Stigler, op. cit.
23 R. Keohane and J. S. Nye, 1972, 1977, op. cit., and numerous articles in *International Organisation* and *World Politics*.
24 In its initial phase, it arose in the work of the Economic Commission for Latin America, particularly that of R. Prebisch. In its elaboration, subsequent authors drew heavily on Marxist terminology, although the mechanistic and somewhat deterministic approach of A. G. Frank and others was much criticised. Among the more articulate writers, it is precisely the search for fundamental alternatives to the present system and for revised policy which situates *dependentistas* in the radical reformist tradition. See, for example F. H. Cardoso, *Politique et Développement dans les Societés Dependentes* Paris: eds. Anthropos, 1971, C. Furtado, *Obstacles to Development in Latin America*, New York: Doubleday, 1971 and O. Sunkel, 'The Development of Development Thinking' in J. Villamil, ed. *Transnational Capitalism and National Development*, Sussex, England: Harvester Press, 1979. An excellent review article may be found in G. Palma, 'Dependency: A Formal Theory of Underdevelopment or a Methodology for the Methodology of Concrete Situations of Underdevelopment'? *World Development*, Vol. 6, No. 7/8, July/August 1978.
25 D. A. Baldwin, 'Power Analysis and World Politics: New Trends vs. Old Tendencies', *World Politics*, Vol. XXXI, No. 2, 1979, and J. Hart, 'Three Approaches to Power in International Relations', *International Organisation*, XXX, Spring 1976.
26 R. Rosencrance, *et. al.* 'Whither Interdependence' *International Organisation*, Vol. XXXI, No. 3, Summer, 1977, p. 441.
27 B. Gosovic and J. Ruggie, 'On the Creation of a New International Economic Order: Issue Linkage and the Seventh Special Session of the UN General Assembly', *International Organisation*, Vol. XXX, No. 2, Spring 1976, p. 344.
28 J. Hart, op. cit. pp. 289–290.
29 K. J. Holsti, 'A New International Politics? Diplomacy in Complex Interdependence', *International Organisation*, Vol. Spring 1978, p. 524.
30 R. Keohane and J. Nye (1977) op. cit. p. 11.
31 Waltz argues that broadly speaking industrial powers are neither vulnerable nor dependent. 'The Myth of National Interdependence', in C. Kindleberger, ed. *The International Corporation*, Cambridge, Mass: MIT Press, 1970, p. 212.
32 The World Administrative Radio Conference, September–November, 1979 organised under the auspices of the ITU, attempted to grapple with the problems of spectrum allocation for the first time in twenty years, including frequencies important to broadcast and private communication carriers.
33 The professional team backing the US negotiations in the last GATT numbered about 160, and that preparing the US position for the WARC was conservatively estimated at 930 (including only government employees). The latter was made by the FCC Chairman in 1977, US Senate Foreign Relations Committee, 'The New World Information Order', November 1977.
34 Examples of this particularly acute political problem, and especially the

hiatus between declared demands and coherent negotiating power, may be found in recent UNCTAD negotiations and the Lomé negotiations between the ACP states and the EEC.

35 An important feature of multilateral negotiation is information handling, reflected in the complexity of items on the agenda at most international conferences. G. R. Winham, 'The Mediation of Multilateral Negotiations', *Journal of World Trade Law*, Vol. 13, No. 3, June 1979, p. 199.

36 J. McHale, op.cit. p. 106.

37 F. J. Bergsten, 'International Economics and International Politics: A Framework for Analysis', *International Organisation*, Vol. XXIX, No. 1, Winter 1975, pp. 20–21.

38 Z. Brzezinski calls for a bloc of developed nations to deal with technological break-throughs in a negotiated manner. *Between Two Ages: America's Role in the Technotronic Era*, London: Penguin Books, 1976, p. 308.

39 See Glen O. Robinson, ed. *Communications for Tomorrow: Policy Perspectives for the 1980s*, New York: Praeger Special Studies and the Aspen Institute, 1978.

40 Report for the US Senate Foreign Relations Committee, 1977, op. cit.

41 John Chippenger, 'Who Gains from Telecommunication Development', unpublished report, Programme on Information Resources Policy, Harvard University, 1976. Details of these issues are summarised in quarterly publication *Transnational Data Report*, Vols 1–2, 1978–9 (Amsterdam).

42 This is what Raymond Vernon has called, 'the Obsolescing Bargain'. *Sovereignty at Bay*, London: Longman, 1971, pp. 46–48.

43 Carlos Diaz Alejandro has called attention to this in 'Finanzas Internacionales: Topicos de Interes Especial para paises en Desarrollo', C. R. Ffrench Davis, and E. Tironi, eds *America Latina y el Nuevo Orden Económico* International, Fondo de Cultura Economica, 1979.

44 As is clear in the review article 'International Bank Lending: A Guided Tour through the Data', Federal Reserve Bank of New York, *Quarterly Review*, Autumn, 1978, Vol. 3, No. 3, while industrialised countries produce a wide variety of information on the subject, none of it is produced in the developing countries and it may not therefore serve all their requirements.

45 Better insulated against the charge that its advice was marked by self-interest, the 'triad' also tempered suspicion of national interest by being based in three major countries. First advising Indonesia, later clients included Sri Lanka, Gabon, Ghana and Peru. *Financial Times*, July 17, 1979.

46 See R. Murray, ed. *Multinationals beyond the Market*, Intra-firm Trade and the Control of Transfer Pricing, Sussex: The Harvester Press, 1981.

47 UNIDO, for example, provides a series of Guides to Information Sources in such sectors as the machine tool or fertiliser industry, or industrial quality control. Each gives directories of professional, trade and research organisations, sources of statistics, marketing and other economic data, but it is unclear how many developing countries actually make adequate use of such data.

48 See, for example, Paul A. Strassman, 'Managing the costs of Information', *Harvard Business Review*, September–October, 1976.

49 These information problems can also directly affect the conduct of foreign policy. See, for example, Basil A. Ince, 'The Information Gap and Non-Consultation: Effects on Foreign Policy-Making in Commonwealth Caribbean States', *International Journal*, Vol. XXIV, 2, Spring 1979.

50 Suggestions for improvement of the uses of expertise are provided by C. J. Lipton, 'Government Negotiating Techniques and Strategies', p. 8, Centre for Transnational Corporations, UN, New York. An alternative proposal has been made (UNIDO Joint Study on International Industrial Cooperation 1979) to supplement expertise available only in the private market by the formation of a Consultancy Agency, financed on a downward sliding scale by the UN initially and later through voluntary contributions. Nothing to date has come of this.

An immediate practical consideration might be a directory of specialised consultants in different fields of negotiation, a register might be updated and verified by relevant intergovernmental agencies. Such directories are already available in the US for domestic purposes.

51 Information in a one-off context must generally be differentiated from the type which is stockable for re-use. Its value when stocked is directly related to local infrastructural capacity for retrieval and use. Information usually has greater value if it is location – or case – specific rather than centrally stored and transmitted from that point.

52 There is some reason to suspect that the widespread use of multinational computer systems may aggravate the present tensions that exist between nation states and MNCs. It is unlikely, for example, that multinational management information systems could be made effective without contributing to the homegenization of problem-solving behaviours, cultural values and public attitudes on a worldwide basis. Moreover, multinational computer systems are likely to enhance the power and influence of multinational organisations whose interests transcend national ones.

B. Nanus 'Business, Government and the Multinational Computer' *Colombia Journal of World Business*, Spring 1978, p. 24.

53 An excellent OECD publication on business information gives a thorough idea of the sources and serivces of data banks available to private enterprise in industrialised countries. The enormous gap between these types of services and the, as yet, rudimentary organisation of information available to LDC governments, state corporations or private firms is apparent. *New Structures and Strategies for Business Information*, OECD, November 1980.

54 Nanus, op. cit. p. 21.

55 The most recent of these is the satellite business systems which have been developed by consortia in the United States following deregulation legislation and a reconsideration of anti-trust laws.

56 A. G. Oettinger, op. cit., Introduction and Z. Brzezinski, op. cit.

57 G. R. Pipe, and A. A. M. Veenhuis eds. *National Planning for Informatics in Developing Countries*, Mouton, 1976.

58 'Informatics' is defined as the rational and systematic application of information to economic, social and political problems. Intergovernmental Bureau of Informantics, *Informatics in the Service of the New International Economic Order*, Rome, July 1978.

59 IBI has among its more than one-hundred member nations many develop-

ing countries, of course, and certain European countries; the US has not joined. Its next large international meeting is to be held in Havana, Cuba in 1983, which some fear may be a largely political affair.

60 S. Ramphal, 'Not by Unity Alone: the Case for Third World Organisation', *Third World Quarterly* (London), Vol. 1, No. 3, July 1979.

# 2

# Uncertainty, Information and the Economic Interests of the Developing Countries

G. K. Helleiner

## Introduction

The 'economics of information' has achieved 'takeoff' within the last decade. Theoretical treatises proliferate on the previously neglected implications of informational imperfections and asymmetries on the functioning of markets, competitive or otherwise. This recent work has been additional to that associated with the older literature on game theory, bargaining, and negotiation, in which information and misinformation play important roles. It has wrought a potential revolution in the 'conventional wisdom' concerning market economies:

> . . . the traditional paradigms of competitive markets, with perfect information and markets equilibrated by the mythical Walrasian auctioneer, are not only not directly applicable, but may be seriously misleading.[1]

> . . . for at least certain important situations, the conventional full information paradigm is not even a good approximation; markets with even a little bit of imperfect information may look distinctly different from markets with perfect information. At the same time, there are undoubtedly situations where the traditional theory will suffice as a good approximation. The relative domains of the alternative theories remain a subject for discussion.[2]

> . . . clearly the traditional presumption of *caveat emptor* has no basis within welfare economics, when information is costly.[3]

At the same time there has emerged a technological revolution in the realm of information storage, retrieval, transmission and processing. New low-cost electronic information processing systems and telecommunications networks have already vastly changed the information scene as far as both market and non-market transactions around the world are concerned; and their full social and economic impact has yet to be felt.

Transnational corporations are among the most important users of new information technology. Much of their use of these systems can be interpreted as the replacement of markets – with their informational and other imperfections, only now being fully comprehended in terms of pure economic theory – with more efficient non-market relationships.[4] Non-market information flows now take place across international boundaries without their notation or enumeration; since statistics for trans-border data flows are still highly imperfect it is impossible to say more than that these flows are already substantial and are rising rapidly. So are the rather more 'visible' associated international transactions in information services, including consulting, data processing, software packages, data access, etc. Governments and international organisations are struggling to keep abreast of these rapid developments and to puzzle out their implications and potentialities.

While the debate on a code of conduct for technology contracts has gone on for many years, international trade in information and information-related services has not, until quite recently, received the same degree of attention, at least from economists. There is a great deal of discussion these days, however, – particularly in the United States – about international trade in services and the need for an agreed international regime (in the US view, freedom) for it. Information and information services are rapidly becoming a matter for high foreign policy.

In recent discussions on transborder data flow, it has become evident that not only are these particular flows unrecorded – certainly not in the balance of payments – but it is not even evident on what basis one might do so. Should they be recorded only when payment is actually made for them? If so, this would be a breach of the conventional practice in respect of goods trade. How should they be valued, if at all? How should intra-firm data flows be treated? At present it seems that only 'services' related to information are recorded in the balance of payments at all, and then quite imperfectly. It is likely that many intra-firm services are left totally out of the accounts since the firm does not always conceive of itself as trading (within itself) the services of its own employees.

The implications of *either* the new theoretical understanding of the role of information in markets *or* the revolution in information technologies, alone, would justify some major effort to analyse afresh the developing countries' interests in the global economic and political system. Together, these changes both in perceptions and in realities present them with an enormous and complex challenge. What *do* we know now about the role of information in the markets, negotiations and other transactions of greatest interest to the informationally weakest in the world economy – the developing countries and particularly individuals, firms and governments within them? What *do* we now need to know about structures of information flows? What *are* developing countries' interests and possibilities in the new information technologies and the policy debates on transborder data flows, etc. to which they give rise? What is likely to be the most efficient approach to policy-relevant research – detailed analyses of specific cases (markets, firms, institutions, countries, etc.)' or a search for more generally applicable principles and objectives' or both?

This chapter is a general consideration of certain of these complex issues, raised initially in the article which provides the introduction to this volume. It is divided into two parts. The first addresses the question, what are exactly the 'problems' of informational deficiency or asymmetry? The second offers some thoughts on possible policy approaches to ameliorate these problems.

## Information 'problems' and the developing countries

What exactly are the 'problems' associated with imperfect information in the effective functioning of markets or hierarchical planning systems? In the rich new literature of the economics of information it is possible to distinguish several different types of informational imperfection. These can be grouped as follows:

(i) uncertainty (incomplete information) about the future;
(ii) 'inherent' uncertainties and deficiencies or asymmetries in informational access about the present, about which there is little that can be done, notably in the realm of product quality;
(iii) informational deficiencies or asymmetries about the present which are 'unnecessary' in the sense that it is possible, at a price, to remove or at least reduce them by 'searching' and/or 'advertising'.

### *Uncertainty about the future*

'Expectations' are formed on the basis of the information available in the present and recent past and its processing according to different rules and with varying degrees of efficiency. While everyone suffers, it is *not* the case that everyone necessarily suffers equally, since different 'expectors' possess different access to data bases and varying capacities to process their contents. A recently popular strand of economic modelling attempts to reduce the importance of such uncertainty (and its differential impact upon different actors) by assuming that all economic actors possess and employ rationally all available and relevant information. In consequence, the price system can be assumed to incorporate efficiently all available information. 'Rational expectations', then, are seen as those based, in one way or another, upon prices.

But even these crude modellers have been forced, in their efforts to bend reality to fit their models, to grapple with the possibility of lags in the conversion of information into price signals and the facts of substantial 'noise' (random error) and frequent 'innovations' (unforeseen events). Keynes pointed out that despite being flimsy, without solid foundation, and subject to sudden and violent change in order to make decisions in an uncertain world, present conditions and majority opinion are generally assumed to be a serviceable guide to the future.[5] Decision-makers in 'remote' parts of the world may not have access even to such rudimentary guides as information on present prices, let alone other conditions or opinions. In volatile commodity, financial and foreign exchange markets the effects of even slight delays in the receipt of information may be disastrous.

Evidence concerning the capacities of specialists to forecast such market phenomena as exchange rates, stock prices, commodity markets, and the like, is mixed at best. One clear conclusion, however, seems to be that some forecasters are a good deal worse than others.[6] Another is that future prices are not particularly good predictors of future spot prices; on the other hand, they are usually at least as good as readily available alternatives.

Whatever the other possible attributes of futures markets they do provide an insurance service to those with fears of unforeseen developments. Risks can obviously be covered via insurance contracts as well. They can also, as will be seen below, be shared by means of appropriate terms of contracts.

Apart from the possibility that 'insiders' or the better-informed may be better predictors of the future, there exists the further possibility that they may, in some circumstances, even be able, through their knowledge, to influence it. Since the future is inherently unknowable, however, there is probably greater and more universal interest in the functioning of futures or contingency markets and the possibility of their creation where they do not exist, than in the perfection of forecasting methods or devices for market manipulation.

## *'Inherent' deficiencies or asymmetries in information*

The new economics of information has been particularly concerned with cases of imperfect information in which prices serve not only to clear markets but also to convey signals as to quality. Quality uncertainty may, to a substantial extent, be 'inherent' in the sense that increased search efforts may uncover very little further information concerning it. In the apparently upside-down world of quality uncertainty, which proves disconcertingly frequently to be the real one, higher prices can generate increased demand, and lower prices can result in lower sales. Buyers may have acquired, through their search activity, virtually complete price information; but, particularly with 'experience' products and 'one-off' purchases, they may still be *unable* to acquire anything like the same degree of completeness of quality information. In such circumstances, low prices may be taken as signals of low quality. Sometimes they are correct signals in that respect. Sometimes, however, they are not; in which case those picking up the incorrect signal are losing the opportunity of a bargain. Low prices may therefore be seen by buyers as involving greater variability of quality – high-risk, high-return opportunities – to be set against the higher prices of more certain products such as, for instance, those associated with larger firms or better-known brands.

These informational problems may directly impede developing countries' entry into world markets in which they have not been selling before. Not only may the quality of their products, particularly manufactured ones but primary goods as well, be regarded by buyers as uncertain, but also the lower are their prices, the greater may be buyer resistance. In the absence of any alternative information, buyers may take lower prices as evidence of greater quality risk.

Imperfect information is also said to generate lender resistance in world credit markets to developing country financial instruments. In such circumstances, higher risk premiums may be totally ineffective in overcoming the problem and, conceivably, could even make it worse. In Eurocurrency markets in recent years this kind of resistance seems to have been overcome to some degree; but the 'herd behaviour' of private banks' lending to particular developing countries suggests again highly imperfect information on the part of bankers and reliance upon signals such as the behaviour of other market participants. So does the recent behaviour of speculative international capital flow.

To some degree, these credit market problems flow from an informational deficiency on the part of the lenders which is exemplary of a wider class of market problems: those of 'moral hazard' and 'adverse selection'.

Insurance contracts and markets, for example, generate incentives for altered behaviour (greater risk-taking or the knowing encouragement of the contingency against which insurance has been obtained) on the part of the insured. This is the well-known 'moral hazard' problem. Since those facing the greatest risks (other things being equal, particularly the degree of risk aversion) are those most interested in insurance, when the same premiums are charged to all insurors there will be 'adverse selection'. An insurance company able to discriminate among parties with varying degrees of risk – that is, one with sufficient information for this purpose – will charge higher premiums to higher-risk insurors. If it cannot do this, it will be faced with adverse selection. Even if, to some degree, it can, it may choose not to offer contracts to those quite willing to pay higher premiums because their very willingness to pay these high rates is evidence of their very high risk and increased moral hazard; thus there may be an upper limit to the price at which the insurance company is willing to contract.

That banks, as has been seen, may not offer credit to willing borrowers above some maximum risk premium above the market rate of interest can now be understood. Similarly, employers may not hire willing workers below some minimum salary, or will oppose or even refuse to go along with efforts to smooth out salary structures. In all these examples there is assumed to be asymmetric information, with the advantage resting with the individual purchaser of insurance, borrower or worker, and the problem is seen as that for the relatively ill-informed firm – deriving from the phenomenon of adverse selection and/or its use in the absence of alternatives, of 'market signals' for the acquisition of its information.

Probably of even greater importance is 'inherent' informational asymmetry of the opposite kind: the relatively ill-informed buyer or seller in the developing countries facing the relatively well-informed businessmen (and governments) of the industrialised countries. The dangers from the provision of misleading or incomplete information in environments within which there are few checks or balances are demonstrated most dramatically in the infamous case of baby formula sales in the less developed countries. In another case, newly TV-orientated Samoan society was found in the mid-1970s to consume enormous quantities of highly advertised Pepto-Bismol 'in a society where no incidence of intestinal or digestive ailments had previously been visible'.[7] There is evidence that in the industrialised countries less educated, lower income and lower status consumers are generally more susceptible to misleading advertising.[8] While firm evidence is lacking, it seems reasonable to hypothesise, with James and Lister, that consumers in less developed countries are likely to form their 'tastes' on the basis of more misinformation than those in developed countries and without the same capacity to process the information coming to them from advertising.[9]

The problems of developing country purchasers are not simply those of easily duped consumers. Governments and firms also face major informational problems as buyers. For developing country purchasers, the informational situation is frequently analogous to Akerlof's classic analysis of the market for 'lemons'.[10] This is particularly the case in the markets for differentiated products like machinery, consulting services, turn-key projects, and the like. Those who sell high quality products are reluctant to sell in the open world market at some kind of 'average price' which includes that for low quality products. The seller of superior goods and services *knows* that his are superior and will not be adequately valued in a market in which 'lemons' are also to be found, and in which the inevitably less well-informed are unable easily to tell the difference. The cautious (or informed) buyer is willing to pay higher prices for higher quality and seeks to avoid the risk of the 'lemon' market in which the good and bad are interspersed. There are some signs – firms' reputations, brand names, etc. – to guide the inherently or otherwise ill-informed; but they typically involve oligopoly 'rents' which poor purchasers legitimately seek to avoid. The poorest and the least informed tend to look for bargains on the 'used car markets' of the world; and they, like used car buyers, are frequently 'ripped off'. Quality uncertainty is central to their problems. The low quality goods (and businessmen) drive out the high quality ones on open markets. Not only, then, are there direct costs to the cheated purchasers from quality misrepresentation by sellers but, more importantly, there are social costs deriving from the driving of some legitimate business out of existence.

Sellers of goods clearly can have an interest in *restricting* information by concealing defects in that which they are selling. Competitors will equally clearly gain from exposing such deficiencies and, to some degree, where there is competition the dangers of incomplete information distorting the selection processes of the buyers may be minimised. Similar considerations obviously arise in bargaining situations, including those relating to essentially political as well as economic issues.

Information deficiencies on the consumers' end of market relationships have generated considerable interest both in the courts and, in the 'consumer movement'. The courts in the US and UK have recognised 'unfairness' and 'unequal bargaining power' as grounds for declaring unenforceable or illegal certain contractual provisions which have been voluntarily entered into, although followers of traditional economic doctrine are uncomfortable with the concept of 'unfairness' in voluntary exchange and frequently find reasons for questioning such legal approaches.[11]

In a world of effectively functioning and competitive markets the selling forms would be disciplined by the market to be truthful in the information they supply and to offer prices which are commensurate with quality. But if markets provide this kind of textbook discipline they do so only 'in the large'; it would be folly to rely upon them for this purpose in specific individual markets or specific transactions. Badly informed consumers may buy inferior products at excessive prices either because of their own inadequate search, because of the inherent difficulty in acquiring information, or because of misleading sales activities on the part of sellers. How much time, effort and expenditure involving search activities is a function of the opportunity cost of the relevant inputs and the expected return from further search activities. The risks of being deceived vary with the nature of the transaction, the reputation of the seller, the type of product being purchased, etc. 'One-off' transactions are riskier than those which are expected to lead to further purchases; purchases from small unknown firms are riskier than those from 'established' firms with reputations to protect; standardised products are less risky to the purchaser than complex or highly differentiated ones.

The smaller the economy, other things being equal, the greater the likelihood of 'one-off' transactions, the longer it will take to 'learn' about the relative merits of different products, and the less information can buyers generally be expected to possess. When small economies are also geographically or communicationally remote, the consumer problems are obviously multipled. If suppliers of information appear to sell information to those who are patently undersupplied with it, similar consumer problems apply to these suppliers as well.

### *Potential response to informational deficiencies or asymmetries*

Both buyers and sellers face information uncertainty. The buyer attempts to reduce his deficiency by search. The seller attempts to provide information by marketing activity, which consists essentially of the collection, use, and selective supply of information. Those who provide information, through advertising for instance, may be able to charge higher prices for identical products because they know that they are thereby reducing their buyers' search costs. A conscientious searcher can often therefore avail himself of cheaper prices. Price dispersion can exist because of imperfect arbitrage which is itself the result of search costs. The advantage taken by a seller of an ignorant buyer would be at a maximum when he perfectly discriminates among each of his buyers, charging each that price which he can, on the basis of his unique demand elasticity (which is in part a function of his degree of ignorance);[12] this is presumably easier to justify when the product being sold is a complex and differentiated one, such as a used automobile.

Some transactors perform much 'better' in markets or in hierarchies or in politics than others. This may be because of differences in preferences or differences in abilities. Differences in abilities may derive from differential costs of becoming informed. While it will generally be agreed that greater search activity will always generate more information, and thereby probably improve the quality of decision-making and/or the confidence in these decisions, it does not necessarily follow that investments in search are always 'profitable'. In the first place the (marginal) cost of search must be taken into account. Secondly, the marginal return from increased information may be small, for instance the extra information gleaned through expanded search may be too small to justify any change in the information possessor's actual behaviour.

There are many ways of acquiring information. Some is 'emergent' and can be acquired simply through waiting, although by the time one has waited it is sometimes too late to employ it productively. Some may be acquired through international non-market sources; and some may be purchased. The market transactions in information give rise to the same sorts of questions as have arisen in the discussion of markets for technology.

If information were freely available to all no one would pay for it. If prices fully reflect all available (at least publicly available) information, then there would be little point in paying for it either. The existence of private specialised information companies, even in highly competitive markets such as those for foreign exchange, suggest that 'insiders' *can* acquire extra information (at a price which is less than the marginal return from it). Barriers to entry may make it possible for high returns to be earned by those who produce it under monopolistic or oligopolistic conditions; these barriers could include scale economies, its under-pricing by firms that sell other services as well.[13]

Information may be offered on direct sale terms; it may be either project packaged or process packaged. As in the case of technology, we still know remarkably little about the channels through which information and information-related services actually flow, or their relative costs and advantages or the potential for better channels in terms of cost, quality or other criteria. And, as in the case of the technology discussion, we do not as yet know much about the buildup of indigenous information systems in practice or in potential.

## Approaches to information policy for the developing countries

Where there is a high degree of uncertainty either because of limited access to potentially available information or because of inherent uncertainty (e.g. about the future) there is an evident extra need for flexibility and the capacity to switch from one type of activity to another. Similarly, where there is an unusually high degree of risk aversion, which characterises those individuals, firms or countries at the margin of survival (physical, economic or political), it may be worth paying a high premium to reduce the potentially disastrous effects of 'surprises'. Among the unfortunate ironies of the Third World is its typical rigidity of production and decision-making structures in the face of the unusually high premium, because of limited information and high risk aversion, on flexibility. Where the avoidable costs of 'switching' are high it may pay to invest considerable sums in improved information acquisition. These circumstances also explain the keen interest in developing countries in international 'buffering' schemes to reduce their uncertainty concerning commodity prices, foreign exchange earnings, terms of trade, exchange rates, etc.

Information-related disadvantages of developing countries in international economic relationships can be addressed in different ways. First, there can be efforts to alter the legal and institutional infrastructure of international exchange through the introduction of codes, guidelines, agencies, etc. which relate to the provision of increased information by informed transactors, generally increased transparency, and at least a minimum of monitoring and enforcement by disinterested parties. Secondly, there can be increased use of contracts in which the risks are shared more equitably than they are when some transactors possess more information than others. Thirdly, and more fundamentally, there can be efforts directly to increase information and information services within the developing countries themselves. These alternatives will be addressed in turn.

### *Improved international legal and institutional infrastructures*

Within nations there are frequently 'consumer protection' laws of various kinds. These, like antitrust laws which are likewise designed to 'soften' the functioning of 'free markets', tend to be both more developed and more enforceable in the developed countries than in the less developed countries. Similar laws or the principles lying behind them might be applied either within those less developed countries with the capacity to enforce them or at the international level (probably through the device of nonbinding 'codes') or both.

Among the informational devices available for the purposes of consumer protection are:

- laws or codes requiring truth in the provision of information by the seller (or, put the other way, the prohibition of false claims);
- laws or codes requiring the provision of more complete information ('disclosure'), e.g. warnings as to health hazards, the establishment of uniform and easily understood measures of particular product characteristics (a 'metric');
- the prohibition of some kinds of information or claims;
- the provision at public expense of increased information about product attributes, prices, etc.

Such remedies of informational deficiencies which operate in any of these ways to 'improve' the flow of information to purchasers (or for that matter to sellers) obviously depend upon effective communication: 'uncomprehended information that is ignored produces no benefits'.[14] In the case of the less developed countries 'effective communication' may be particularly difficult. Even in the American context, it has been observed that messages 'must be consonant with the information processing capabilities of the target audience . . . and must consider the limitations of the medium in which it will be placed'.[15]

In this area, as in others, it is possible that developing countries might seek exemption from 'disclosure' requirements; but, as far as basic standards are concerned this would not probably be in their interest. (This is not to say that they would not gain from the restriction of some other kinds of information such as that on crop prospects.)

Whether any of these sorts of measures could be agreed at the international level or, if agreed, actually enforced, is an open question. Certainly in the foreseeable future any such schemes seem rather unlikely. On the other hand, gross violations of accepted Northern standards in advertising and marketing on the part of Northern-based transnationals (if perhaps not on the part of local firms) may be susceptible to 'control' by Northern public opinion, if the Nestlé case is any kind of a harbinger of future prospects.

### *Altered distribution of risks in contracts*

Informational deficiencies may also be attacked via policies not dealing directly with information at all, e.g. the regulation of product standards and/or prices, the regulation of the terms of contracts (e.g. through the requirement of warranties), price stabilisation, etc. Broadly speaking, this second category of policy alternatives seeks to alter the distribution of risks with respect both to the unknowable future and to quality dispersion, information on which may be asymmetrically available.

Market transactions are usually assumed to be 'non-contingent', in the sense that an agreement is made to buy or sell at a point in time or over a period, regardless of what new circumstances may develop in the interim, or of subsequent satisfaction with the original contract. The transaction costs involved in detailed provisions relating to all manner of future contingencies create an obvious disincentive to their use. But the presence or absence of such provisions has clear implications for the carrying of risk.[16] 'Non-contingent' terms of contracts may include implicit provision for risk within the conventionally-defined 'price', e.g. risk premiums charged on loans to 'less creditworthy' borrowers.

Political bargains are more often than economic contracts of a contingent character: 'if these events occur, then these arrangements and reactions will follow'. Where longer-term or particularly comprehensive treaties or agreements are negotiated, there is frequent resort, however, to an 'escape clause' (or several of them) so as to preserve the legalities while permitting contingent observance of their terms:

> . . . complete, fully contingent, costlessly enforceable contracts are not possible . . . Rather than the impersonal marketplace of costlessly enforceable contracts represented in standard economic analysis, individuals in most real world transactions are concerned with the possibility of breach and hence the identity and reputation of those with whom they deal.[17]

Since the possible contingencies are so many and since it is often difficult to prove a breach of contract, a common option for transactors is simply to 'renege on the transaction by holding up the other party, in the sense of taking advantage of unspecified or unenforceable elements of the contractual relationship'.[18] If this prospect is understood in advance, again, the terms of a contract may implicitly reflect it, even if they appear to be 'non-contingent'.

A conscious effort could be made to reduce risks for more informationally deficient transactors through encouraging the reduction of uncertainty surrounding the eventual effects of contracts. This could be done by greater use of Bisque clauses in financial contracts, explicit inclusion of contingency arrangements or performance bonds in commercial contracts, the use of basket currencies for the denomination of contracts, and the like. There is obviously a considerable history of efforts to stabilise certain commodity prices and, more recently, to offset fluctuations in export earnings and even some of those in import requirements. In all such schemes, however, there remains the question of who is to bear their cost and ultimately, therefore, the question of whether improvements will be truly 'additional'. As has been seen, 'better' terms of contracts in the sense of reducing uncertainties may be offset by worsened prices. Increased compensatory financing may be at the expense of some other kind of finance for generally the same set of beneficiary countries.

### *The development of indigenous information systems*

A third policy set is much more fundamental: measures designed directly to improve the information and the information handling capacity of the informationally most deficient. These can be undertaken by the international community; by groups of countries allied by regional, sectoral, functional, political or other interests; or by individual national governments or firms. They can relate to forecasting, quality and price dispersion, and a wide variety of other types of information. Transnational corporations may play an important positive role in this area. And it is in this area that the new information technologies, if they can be appropriately and effectively harnessed, may be of great significance.

Scale economies in the collection, processing, storage and transmission of information have always been at the centre of discussions about informational adequacy on the part of economic and political actors. The relative disadvantages of small and medium-sized firms in world markets – whether importing or exporting – have long been attributed in large part to their relative weakness in information. To overcome this weakness the traditional remedies have been the formation of large-scale trading houses (or marketing boards), the use of the information networks of the international financial institutions with which small actors are associated, the purchase of information from specialist firms in the relevant sectors, etc. Governments have frequently tried to assist informationally weaker national firms in international trade through the services of commercial attachés, trade missions, export promotion bureaux, statistical services and the like. Above all, if scale economies are important, is the need for cooperation among small actors for the purpose of reducing cost. This is evidently an area for major potential gain, particularly for the smaller countries, through economic and technical cooperation with other developing (or even developed) countries.

## Conclusion

According to some, the revolution in telecommunications and informatics is generating a potential 'sea change' in the relative informational advantages of small and large actors. If there is the potential for virtually immediate access to all manner of market and other data through cheap and reliable satellite communication systems it may no longer be economic in the future for firms to try to develop collaborative or their own (decreasing cost) information networks. The comparative informational advantage will then rest with those with the capacity to process and interpret information; and that may be much less a function of scale than the capacity to develop an information network is. Small-scale 'software' firms should be able to thrive internationally as, to some extent, they already do within the industrialised countries. Opportunities for developing countries may exist in this arena much more than in the hardware portion of the new information sectors: India, for instance, was exporting informational software well before it was able to sell hardware competitively abroad. The key questions would then be those of what may be the most useful forms of indigenous or custom-made 'software' for developing countries, and where they are to come from.

In the meantime, the developing countries must obviously work with what is at hand. What are the sensible next steps for developing countries of different economic structures and different levels of overall development? While it may be possible for newly industrialising countries to hire 'senior information systems executives' at $200,000 per year (advertisement in *The Economist*, July 18, 1981), many other developing countries are not so fortunate. It seems to me that in many countries the issues have as yet scarcely been posed.

The uncertain decision-maker seeking to make the best use of his restricted amount of information is highly dependent upon his own 'communication system'.[19] What information does he most need? Where does he now get his information? What is its quality? What does he pay for it? What measurable difference does it ultimately make to him?

To get at these issues, and thus to be able to consider improved alternatives, one must know much more about global information channels; the technology they employ; their degree of centralisation and market structure characteristics; the possible biases in them; their costs, and the principal factors influencing them; the capacities of users to employ what they channel; and the gaps in what they offer in relation to already perceived needs, etc. These are empirical questions which are likely to vary among sectors, countries, firms and time periods; and they are likely to be changing rapidly. The structures of current information flow for use by developing countries, whether through markets, or other institutions, present a potentially rich field for social scientific research. Those who can employ such research to understand and eventually to help to shape a more equitable use of information in international transactions, will make a truly rich contribution.

## Notes

1 Joseph Stiglitz, 'Equilibrium in Product Markets with Imperfect Information', *American Economic Review*, Vol. 69, No. 2, May 1979, p. 344.

2 Joseph Stiglitz, 'Introduction', *Review of Economic Studies, Symposium on Economics of Information*, Vol. XLIV, No. 3, October 1977, p. 391.

3 Joseph Stiglitz, 1979, op. cit., p. 344.

4 Oliver F. Williamson, *Markets and Hierarchies, Analysis and Antitrust Implications: A Study in the Economics of Internal Organization* (The Free Press, New York), 1975.

5 John Maynard Keynes, *The General Theory of Employment, Interest and Money*, 1936, pp. 147–64.

6 See, for example, Richard Levich, 'Analyzing the Accuracy of Foreign Exchange Advisory Services: Theory and Evidence', and 'Comment' by Michael Dubin, in Richard M. Levich and Clas G. Wihlborg (eds) *Exchange Risk and Exposure, Current Developments in International Finance Management* (Lexington, D. C. Heath), 1980.

7 From a review in *The Economist* of Wilbur Schramm *et al.*, *Bold Experiment: The Story of Educational Television in American Samoa* (Stanford University Press), 1981.

8 Jeffrey James and Stephen Lister, 'Galbraith Revisited: Advertising in Non-Affluent Societies', *World Development*, Vol. 8, No. 1, January 1980, p. 93.

9 Ibid. p. 94.

10 George A. Akerlof, 'The Market for "Lemons": Quality Uncertainty and the Market Mechanism', *Quarterly Journal of Economics*, Vol. 89, August 1970.

11 Benjamin Klein, 'Transaction Cost Determinants of "Unfair" Contractual Arrangements', *American Economic Review*, Vol. 70, No. 2, May 1980, pp. 356–62.

12 Steven Salop, 'The Noisy Monopolist: Imperfect Information, Price Dis-

persion and Price Discrimination', *Review of Economic Studies, Symposium on Economics of Information*, Vol. XLIV, No. 3, October 1977.

13 Steven Salop and Joseph Stiglitz, 'Bargains and Ripoffs: A Model of Monopolistically Competitive Price Dispersion', *Review of Economic Studies*, *Symposium on Economics of Information*, Vol. XLIV, No. 3, October 1977.

14 Howard Beales, Richard Craswell and Steven Salop, 'Information Remedies for Consumer Protection', *American Economic Review*, Vol. 71, No. 2, May 1981, p. 411.

15 Ibid. p. 413.

16 Geoffrey Heal, 'Guarantees and Risk-sharing', *Review of Economic Studies*, *Symposium on Economics of Information*, Vol. XLIV, No. 3, October 1977, and
Michael Spence, 'Consumer Misperceptions, Product Failure and Producer Liability', *Review of Economic Studies*, *Symposium on Economics of Information*, Vol. XLIV, No. 3, October 1977.

17 Klein, op. cit., p. 356.

18 Ibid.

19 M. Shubik, 'Information, Rationality and Free Choice in a Future Democratic Society' in D. M. Lamberton (ed.), *Economics of Information and Knowledge* (Penguin, London), 1971.

# 3

# Transnational Corporations and Transborder Data Flow*

## Introduction

Over the past decade, rapid advances in both computer and telecommunication technologies have led to their convergence into a new activity, telematics. Modern telecommunication facilities have overcome time and distance as major obstacles to the access of sophisticated computer services for the processing, storage, and retrieval of machine-readable data. The transnationalisation of this process, in turn, has given rise to transborder data flows, which are international data transmissions over transnational computer-communication systems. Transnational corporations are of key importance to this new activity. For them, transborder data flows are both a commercial good as well as a management tool.

More specifically, transnational corporations play a central role in establishing the required telecommunications infrastructure; they produce the necessary computer hardware, software, and peripheral equipment; and they offer a growing range of data-processing services as well as access to an expanding number of machine-readable data. In the process, an international market for transborder-data-flow equipment, software and technology as well as for the production, processing, transport, and distribution of a wide variety of data of economic, political, social, and legal interest has been created. In addition, transnational corporations utilise computer-communication systems for the rationalisation of world-wide corporate management. As a result, transnational corporations have improved their capacities to manage

* This chapter was originally issued as United Nations document E/C.10/87 and Corr. 1, under the title, *Strengthening the Negotiating Capacity of Developing Countries: Transnational Corporations and Transborder Data Flows. An Overview.* It is based on a technical paper, entitled *Transnational Corporations and Transborder Data Flows: A Technical Paper* (ST/CTC/23) which also contains the sources for the data used here as well as an elaboration of the various points summarised in this paper. It was produced by a member of staff at the Centre on Transnational Corporations in New York.

and co-ordinate more efficiently their global operations. To date, however, both access to the international data market as well as the utilisation of the underlying technology and equipment for various purposes have remained almost exclusively confined to institutions of the developed market economies. The experience of developing countries (which reflects, of course, their level of development) has been mostly one of supplying data, buying the processed data (i.e. information), and acquiring the capital goods needed for their essentially rudimentary telematics sectors.

The significance of these developments lies in the importance of appropriately applied accurate domestic and foreign information as a central factor for decision-making, including all matters related to development. Information is a precondition for identifying alternatives, reducing uncertainty about their implications, and facilitating their implementation. As such, it is also a critical resource for enhancing bargaining strength in the pursuit of clearly defined purposes.

In the light of the importance of information, the present structure of the international data market and the uneven distribution of the capabilities of applying transborder data flows raise a number of concerns. For the developing countries in particular, they include the implications of transborder data flows for the international division of labour; the capacity of the developing countries to establish their own telematics and transborder-data-flow facilities; the competitiveness of domestic corporations of host countries; the bargaining position of host countries *vis-à-vis* transnational corporations and, more generally, the developed countries; and even issues relating to national sovereignty. Of course, the nature of any possible impact may be a function of the level of development. Moreover, care has to be taken to distinguish the specific impact of transborder data flows from the general effects of automation and telematics; these effects are not being considered here.

The implications of concerns about transborder data flows reach beyond the economic sphere and may affect the relations between developed and developing countries in general. For, to the extent that information is a basis of power, the ability to collect, store, access, process, and work with it can give countries (as well as groups and institutions within them) political, economic and social advantages over other countries (and their groups and institutions). Transborder data flows may thus lead to an accentuation of existing imbalances – and, in fact, the emergence of a new set of imbalances – between the developed and developing countries.

On the other hand, if the developing countries' capabilities to utilize transborder data flows and their access to the international data market can be improved sufficiently, and if these flows allow applications that are useful for developing countries, these flows may become an instrument for the advancement of the development objectives of the developing countries. In fact, under certain circumstances, transborder data flows may contribute to the establishment of an informational infrastructure for a more self-reliant development in the framework of greater economic co-operation among developing countries. A key issue is, therefore, how it can be ensured – through the provision of adequate access to the international data market and through the strengthening of appropriate national industrial and guidance capacities – that the entire international community can participate in the opportunities created by transborder data flows.

## Transborder data flows

An important factor in the economic development during the past decade has been the rapid development of technologies relating to computers and telecommunications, providing for cost-effective and sophisticated data-processing, storage and retrieval at virtually any location. The importance of this factor may be even greater in the years to come. As a recent study prepared by the Organisation for Economic Co-operation and Development observed: '. . . *the electronics revolution will change the face of advanced industrial societies*. Production, transmission and processing of the most varied information will be at the heart of economic activity and social life . . . ; in fact, 'through its links with data processing and telecommunications, . . . *the electronics complex during the next quarter of a century will be the main pole around which the productive structures of the advanced industrial societies will be reorganised* (author's italics).'[1] If such changes are likely to materialise in the developed countries, it is very probable that the developing countries will not remain insulated from them. Rather, these changes can be expected to transform their economies as well, sooner or later.

Telematics,[2] the merger of computer and telecommunications technologies, is one of the principal driving forces behind these changes. It is, in fact, the precondition for the emergence of transborder data flows.

### *Definition*

Broadly defined, transborder data flows are movements across national boundaries of machine-readable data for processing, storage or retrieval.[3] Such movements can be effected by non-electronic means, e.g. on magnetic tapes, discs, punched cards, or other media. Increasingly, however, electronic means are used, which presuppose the availability of a telecommunications infrastructure. In these cases, transnational computer-communication systems are established, i.e. arrangements whereby one or several sophisticated computers in one or several countries are linked to affiliated computers in other countries and through them (or directly) to remote terminals. The trend in transborder data flows is towards the usage of transnational computer-communication systems.

Data have, of course, moved across national boundaries before the advent of telematics. But traditional media (such as postal, telephone and telex services) are increasingly being replaced by the electronic transmission of data. The fusion of advanced electronic processing capabilities with modern telecommunications facilities and the speed, accessibility, and inter-active capabilities through which time and distance are overcome as obstacles to large-scale information mobility, give transborder data flows their potency. Apart from these technological changes, a number of other factors have also been responsible for the growth of transborder data flows. To begin with, the increasing role of the production and distribution of information goods and services was bound to transcend national boundaries, particularly in a world which is growing increasingly interdependent. Further, the expansion of world trade and the internationalisation of such information-intensive industries as banking, insurance, tourism, and air transport, intensified the need for mechanisms ensuring the instantaneous availability and dissemination of data. Finally, the spread of transnational corporations partly fuelled, and was partly dependent upon, the creation of transnational computer-communication systems that permit the rapid transmission of large volumes of data to control and co-ordinate large and complex organisations with functionally diverse and geographically dispersed operations.

As indicated above, transborder data flows involve the point-to-point delivery of messages. In addition, they are normally of a proprietory nature and are based on contractual relationships between parties. Media products which involve mass diffusion – especially broadcasting and television – are, therefore, excluded from consideration here. This is not meant to imply that the issues associated with mass media do not deserve attention, but rather that they are of a different nature; furthermore, they are being examined by the United Nations Educational, Scientific and Cultural Organisation.[4]

In the following, two categories of transborder data flows are distinguished and described: commercial and corporate data flows. Before this, however, the growing infrastructure of transborder data flows is documented briefly.

## *The infrastructure of transborder data flows*

*Networks* Apart from computers, transborder data flows require an infrastructure through which data transmissions can be made. At the national level, this infrastructure is provided by traditional telephone networks. As a rule, these are common carriers, i.e. authorised entities (often with a monopoly position) that provide universal communication services for a fee. Where countries are contiguous, these common carriers are merely linked up with each other and transborder data transmission can take place. For transoceanic traffic, international record carriers (in most countries the domestic common carriers) are authorised by bilateral agreements to provide interconnections through submarine cables and, increasingly, through satellite links among common carriers. Most of the capacity of these installations is in the northern hemisphere.

The utilisation of satellites for data transmission will further accelerate when direct satellite services specifically designed for corporations become more common. One such system, the Satellite Business Systems, is already in operation and its most important users are transnational corporations. It replaces terrestrial communication systems by direct contact between earth stations located on the premises of the users, satellites, and roof-top antennas. It provides integrated digital data, voice, and image-transmission services for intra-company transactions.

Value-added carriers (carriers that lease lines from common carriers and combine them with computer equipment of their own) further enhance the usefulness of a telephone network for data-transmission purposes. This creates a communication network dedicated solely to data transmission (dedicated network). For many users, value-added networks allow a faster, more reliable, and more cost-effective link-up of computers than could be obtained if users were to lease standard lines from common carriers and were to invest in the necessary equipment themselves.

The most important international value-added networks are currently EURONET, the Nordic Public Data Network (both of which are public data networks) and the privately-owned transnational TELENET and TYMNET (see Figs 3.1 and 3.2). So far, only the last two networks have an intercontinental span. Increasingly, however, networks are becoming interconnected. The most recent example is the establishment of a computer-communication link between Eastern European centrally-planned and Western market-economy countries through the gateway provided by the International Institute for Applied Systems Analysis (see Fig. 3.3). This extension of computer-communication networks to some socialist countries contributes to the rapid growth of an expanding transnational telecommunication infrastructure for transborder data flows. These networks are the basis of an international data market with on-line capabilities.

*Access and usage* Access to computer-communication networks is gained at network-termination points. In Western Europe alone, the number of network-termination points is estimated to increase from 393,000 in 1979 to 1,620,000 in 1987. Over the same period, the number of terminals is expected to rise from 625,000 to nearly 4 million. Forecasts suggest that the increased access to computer-communication networks will lead to an expansion in the volume of data flows in Western Europe from approximately 136 million transactions in 1979 to almost 800 million in 1987; of these, the share of international transactions is expected to rise from 10 to 15 per cent during the same time period.

Two features of these data flows are particularly noteworthy. One is the strong tendency of flows to be oriented towards capital cities and main metropolitan areas, primarily because the principal data-communications users tend to be located there. The other one is that the least advanced (in terms of GDP per capita) countries in Western

**Figure 3.1** *The major value-added regional networks: EURONET and the Nordic Public Data Network, 1981*

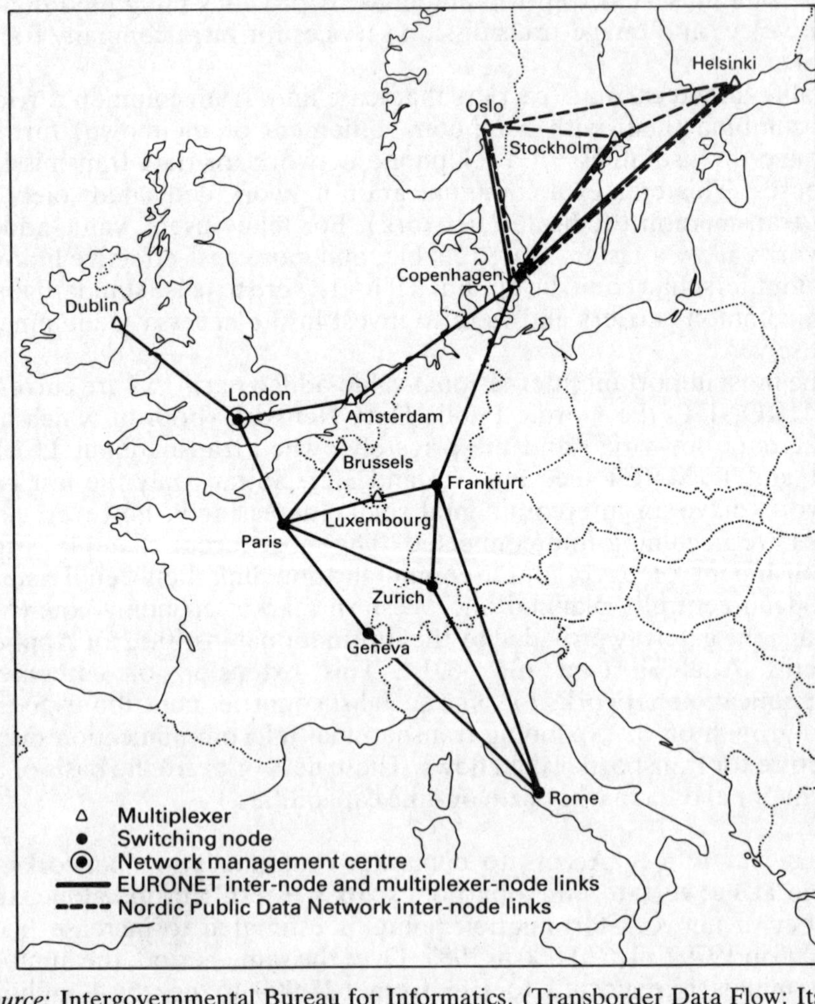


*Source*: Intergovernmental Bureau for Informatics. (Transborder Data Flow: Its Environment and Consequences) (Rome: IBI, 1980), document SPIN 231, p. 22.

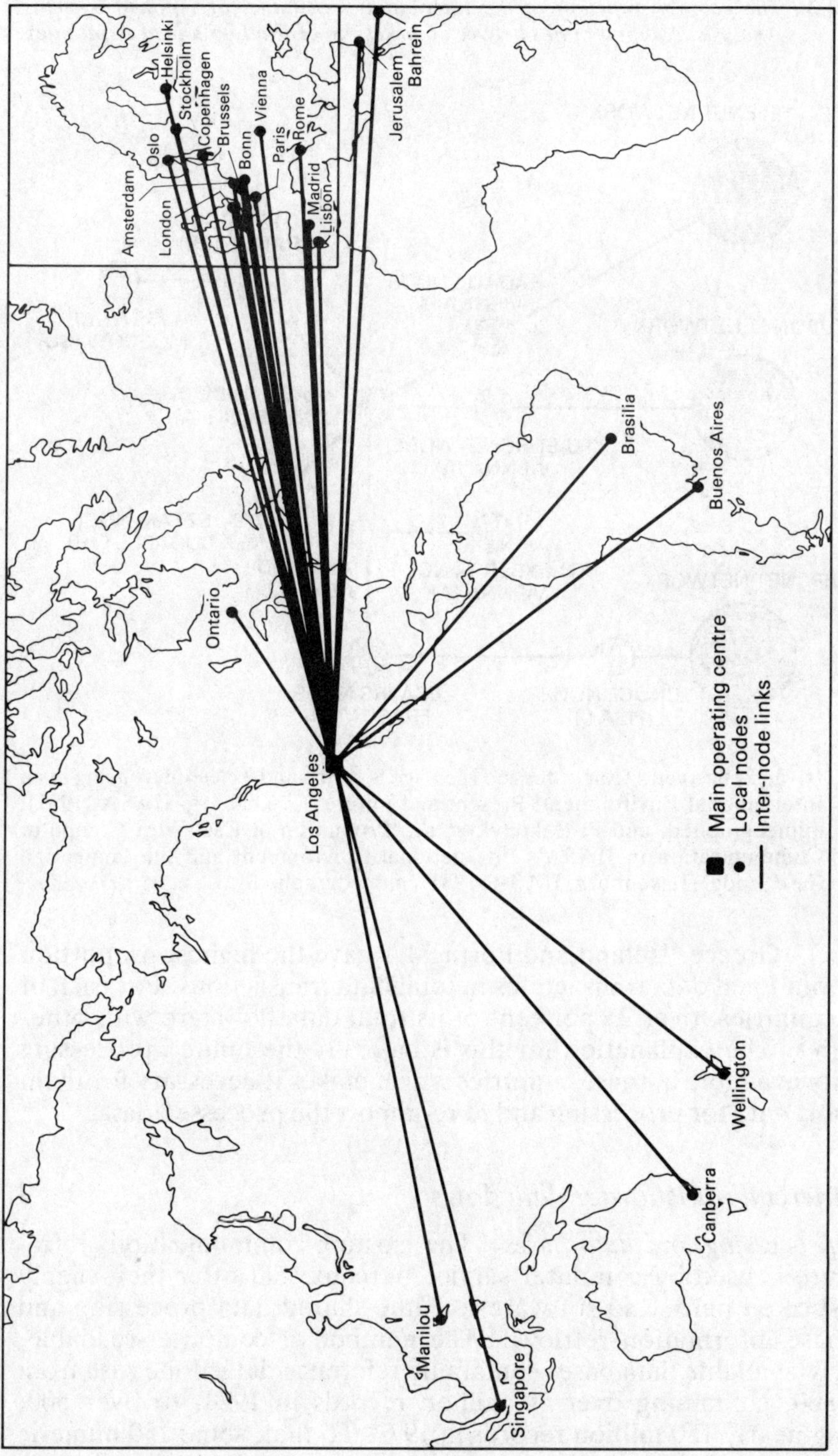


**Figure 3.2** *The major value-added international networks: TYMNET*

**Figure 3.3** *The gateway function of the International Institute for Applied Systems Analysis: computer links between market and centrally-planned economies*

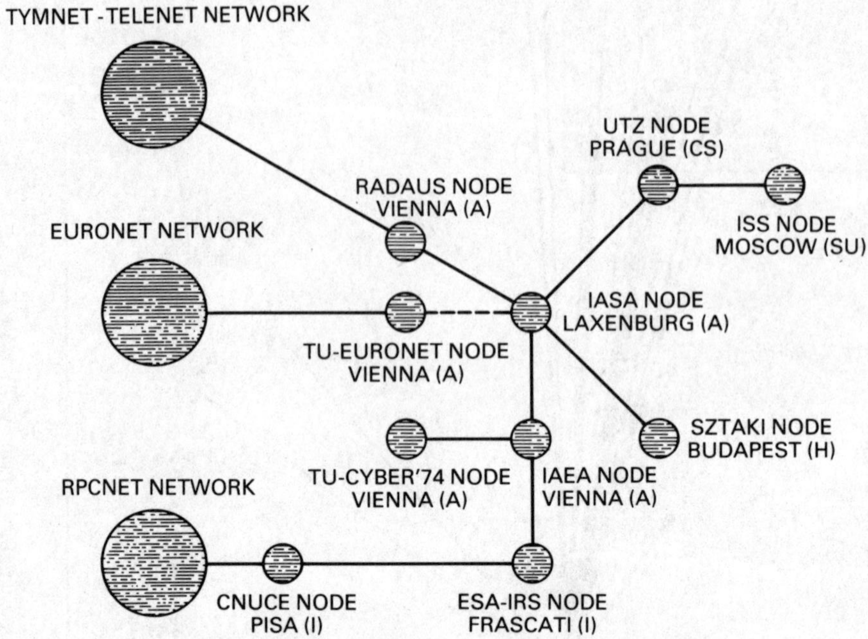


*Source*: Istvan Sebestyen, 'Computerised Message Sending and Teleconferencing in an International Environment: Present and Future' (Laxenburg: IIASA, 1981), mimeographed, and P. Bakonyi, *et al.*, 'Promotion of East-West Computer Communication in IIASA's International Environment and the Hungarian Case Study' (Laxenburg: IIASA, 1981), mimeographed.

Europe – Greece, Ireland and Portugal – have the highest proportion of international data transactions in total data transactions. For each of these countries, over 25 per cent of its total data flows are with other countries. The explanation for this is basically the limited processing capacity available in these countries which makes it necessary for them to export data for processing and to re-import the processed data.

## *Commercial transborder data flows*

*Data processing and data bases* The existing communications infrastructure is used by computer service bureaux that offer their highly sophisticated hard- and software for time-shared data processing and data-base information retrieval. The number of computer-readable, publicly-available data bases containing reference data alone rose from over 300, containing over 50 million records in 1974, to over 500, holding nearly 150 million records in 1979. To that, some 150 numeric data bases have to be added. These data bases contain a wide variety of information pertaining to such matters as financial time-series data on individual corporations, industry data, security sales, foreign-exchange developments, commodity trends, technology available for licensing, patents, and the like.

The expansion of information-retrieval services has contributed to a dramatic expansion of the computer-services market. Total domestic revenues of that industry in the United States alone rose from about $2 billion in 1970 to $8 billion in 1978. In addition, international revenues increased from approximately $300 million in 1972 to $800 million in 1978.

*Remote sensing* A special category of commercial transborder data flows consists of remote-sensing data. Satellites equipped with earth-resource sensors monitor the earth and collect data about its environment and resources and communicate them to the ground. Among the areas in which remote sensing technology is being applied are agriculture, forestry, geology and natural resources, hydrology and water resources, urban planning, and oceanography and marine resources. More concretely, for instance, remote sensing data, if combined with meteorological satellite data as well as information from ground-based weather stations and used in sophisticated computer models, allow crop-production (including crop-health) forecasts well in advance of traditional methods. Computer-assisted intepretation (which is needed, for instance, for crop forecasts) requires a very sophisticated processing technology which only a few service bureaux have developed. Their clientele includes governments and private industry.

Naturally, the users of remote-sensing data are in a better position to prepare themselves for any developments, e.g. by buying and selling primary commodities in the futures market. Similarly, information about geological structures – to date the most important application of remote sensing data – can be important in determining the location of mineral or oil deposits, and hence be of great value to institutions that negotiate about the right for prospecting and developing natural resources.

*Corporate transborder data flows* As costs decreased, the utilisation of the computer spread and data communication networks expanded. The electronic transmission of data became more attractive for business organisations in general and transnational corporations in particular. As one observer noted: 'For the multinational enterprise, the importance of the computer lay in the fact that the routine data needed for the direction and control of global operations could be transmitted in vast quantities and could be retrieved and regrouped with lightning speed.'[5] In fact, most transnational corporations have computer-communication systems under the broad definition of such a system; under the narrow definition (i.e. point-to-point computer communication), an estimated 100–150 transnational corporate computer-communications networks existed in 1978 in Western Europe alone. As the Business and Industry Advisory Committee of the Organisation for Economic Co-operation and Development pointed out, the fact that transnational corporations 'have found it worthwhile to invest in private networks is an indication of how essential well-functioning IIF (International Information Flows) is to modern business.'[6]

Transborder data flows enable transnational corporations to raise the operational efficiency of their integrated global strategy. For some industries, transborder data flows have almost become a prerequisite for the conduct of transnational operations. Most of these flows are intra-company data transactions and involve operational data needed for control, decision-making, conduct of operations, and administration. Some of them, however, are also inter-company flows that utilise dedicated networks established to service the specific needs of their subscribers. In both cases, existing transnational computer-communication systems are utilised or new ones are established.

*Corporate networks* The reasons for which transnational corporations have established computer-communication systems are manifold. Most importantly, transnational corporations with integrated global operations are complex organisations because of the geographical distribution of markets and corporate assets, the diversity of products and services that are offered, the variety of economic, fiscal, legal and political environments that have to be taken into account, and the sheer volume of transactions. To control such an organisation, to manage it efficiently, and to monitor its performance, requires a large amount of information, especially for the parent corporations in which much decision-making power is typically vested. Furthermore, this information has often to be transmitted and processed without delay and must be stored and up-dated continually for future reference. Some of it (as in the case of various reservation systems, credit authorisations, and inventory control) must remain accessible virtually on an on-line basis, while other data (like financial transactions) must be validated immediately.

Changes in the international environment have also played an important role. During the 1950s and 1960s, the international business environment was relatively stable and predictable. During the 1970s, however, uncertainties in this environment increased considerably. Movements in exchange rates and interest rates, for instance, became more frequent. Similarly, price movements in input and output markets developed differently. Corporations wanting to seize upon business opportunities that offered themselves under these changing circumstances – or desirous of avoiding risks and decreasing their vulnerability – needed a monitoring approach that allowed them to allocate their resources swiftly. Finally, the slower growth of production in many sectors made it more difficult to treat foreign affiliates as single profit centres; in other words, the efficiency of the corporate system as a whole had to be enhanced as much as possible.

In line with the growing need for information for the management of transnational corporations, the areas of application of transborder data flows range across most corporate functions: production planning, standardisation and co-ordination; financial control, research and development; customer and purchasing questions; legal reporting; and employee relations. In many of these areas, the data of foreign affiliates can be processed by a computer system servicing the entire corporate network. Data-processing systems allow the establishment of economies of scale (which can be exploited on a time- and hence cost-shared basis), the avoidance of duplication, and the rationalisation of worldwide corporate activities in the interest of over-all better performance. Thus, transborder data flows give transnational corporations a greater capacity to maintain the coherence of the corporate system as a whole and to allocate more efficiently and effectively corporate resources.

*Closed user-group networks* While a number of transnational corporations could probably operate – albeit at a cost – at a lower level of transborder data flows, some industries have become so information-intensive that transborder data flows have evolved into an essential requirement for effective and competitive corporate performance. Many corporations, particularly in service industries, depend on the quick availability and dissemination of information and hence need an elaborate data support system. In some information-intensive industries, it has, therefore, become advantageous, if not imperative, to establish closed user-group transnational computer-communication systems, i.e. networks that service the needs of their subscribers for inter-firm transborder data flows. Three major systems of this kind exist, two of which, Eurex Sa. and the Society for Worldwide Interbank Financial Telecommunication (SWIFT), service the financial community and one, the Société Internationale pour la Télécommunication Aéronautique (SITA), services the airtransport industry (see Fig. 3.4).

*Centralisation and decentralisation* In all corporations that capitalise on the possibilities offered by transborder data flows, a balance has to be found between centralising and decentralising operations. In some areas, like production specialisation, a certain measure of decentralisation may be essential for the optimal performance of foreign affiliates or regional profit centres. Transborder data flows can be a management tool to make a proper centralisation/decentralisation model workable: 'Management will therefore be able to determine which controls and standards should be applied throughout the organisation to facilitate

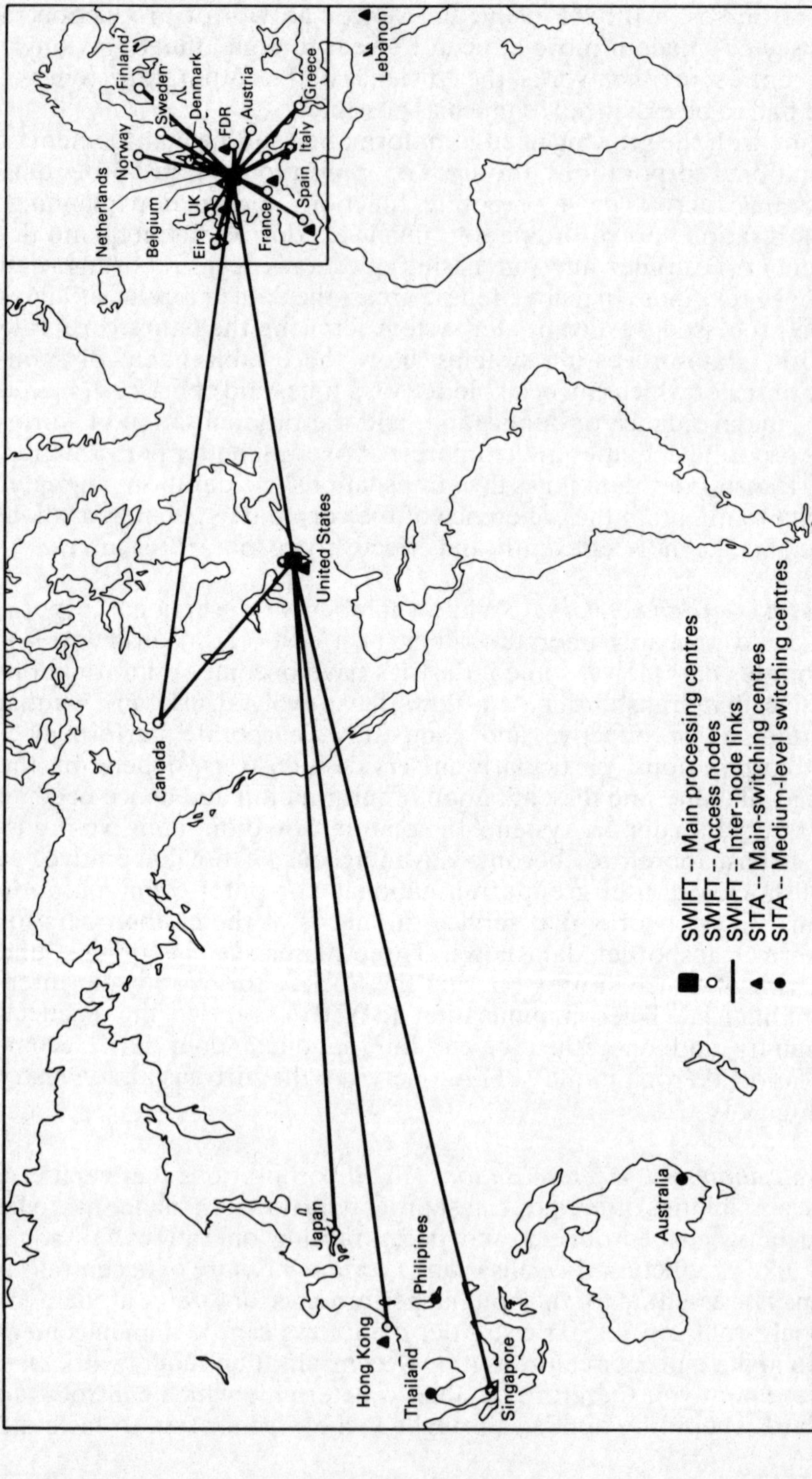


**Figure 3.4** *The major closed user-group networks: SITA and SWIFT*

meaningful corporate control, based on business considerations, rather than the limitations of communication/computing facilities. Planning can be decentralised within meaningful constraints and yet controlled centrally.'[7]

On the other hand, the new technologies may well be used merely to automate routine work without transferring real decision-making authority to affiliates. In any event, even very decentralised corporate organisations require the capacity to co-ordinate and consolidate the different types of data that are pertinent to them. In many situations, the actual requirements for cost-efficient operations, the need to monitor the performance of the entire corporate system, or the tendency to take advantage of existing central installations may favour central co-ordination.

As a result, parent corporations may have a greater potential than ever to exercise control over their entire corporate systems. Yet, even where some balance between centralised and decentralised operations has been found, it may involve a clear division of tasks between the headquarters and affiliates of transnational corporations. The latter tend to send data for processing abroad or engage only in simple processing, while the former – where the large computer facilities are normally located – execute the sophisticated processing and undertake the various tasks associated with the maintenance and development of a computer-communication system as a whole. Thus, the over-all effect may be a greater specialisation of individual foreign affiliates coupled with their stronger integration into the parent system, an upward moving of control within the corporate hierarchy, and the increased utilisation of the central control capabilities that transborder data flows confer upon headquarters management. Conversely, the decision-making autonomy of affiliates and their capacity to respond directly to particular objectives and requirements of host countries may decrease, while the dependence of the various units of a transnational system – headquarters, regional offices, manufacturing sites, sales outlets, etc. – on uninterrupted, relatively cheap, and reliable transborder data flows may increase.

## *Features of the structure of the international data market and some applications of transborder data flows*

The preceding documentation of the infrastructure of transborder data flows and its commercial and corporate applications allow a number of summarising observations about some of the current features of this new phenomenon.

The manufacturing of the computer equipment supporting transborder data flows, the production of the accompanying software, and the development of the underlying technology are limited to the industrialised countries, with transnational corporations based in the United

States playing a leading role. The distribution of computing facilities across geographic regions is uneven, reflecting varying levels of development and of availability of skilled human resources. Developing countries account for only a fraction of the total market. Similarly, most capacities for telecommunication – telephones, telephone networks, submarine cables, satellites – are located in the northern hemisphere of primarily service users there. The distribution of the equipment required for transborder data flows is thus largely limited to the developed countries.

The establishment of transnational computer-communication systems and their commercial use are also concentrated in the developed market economies. Publicly-available communication networks specifically created for the transmission of data now exist or are nearing completion in most of these countries. They are linked up internationally, but have only a few extensions into developing and socialist countries. Data processing undertaken by service bureaux over the existing networks is also mostly confined to the developed market economies, as is the *de facto* access to the expanding network of computerised data bases. The production, processing, transportation and distribution of machine-readable data for commercial use are almost exclusively a phenomenon of the developed market economies. Finally, while the market for remote-sensing data is wider, practically the only supplier of publicly-available remote-sensing data is an agency of the United States Government, their main users are private corporations, and almost all computer-assisted interpretation of such data is limited to a few developed countries.

Most corporate computer-communication systems are structured in such a manner that the main processing facilities are located at the seat of the parent corporation (or at least in the home country), while foreign affiliates, especially those in developing countries, are linked to the system through terminals. Also, the majority of principal switching centres in the closed user-group networks is located in developed market economies.

Thus, the international data market and the application of transborder data flows for corporate purposes are largely limited to the developed market economies. Almost all aspects of the market are a domain of transnational corporations. Developing countries participate primarily as data suppliers, buyers of the finished product, information, and purchasers of the capital goods needed for their own, so far mostly rudimentary telematics sectors. The resulting over-all pattern of transborder data flows – which, of course, may change in the future – is schematically summarised in Fig. 3.5.

**Figure 3.5** *Patterns of transborder data flows*

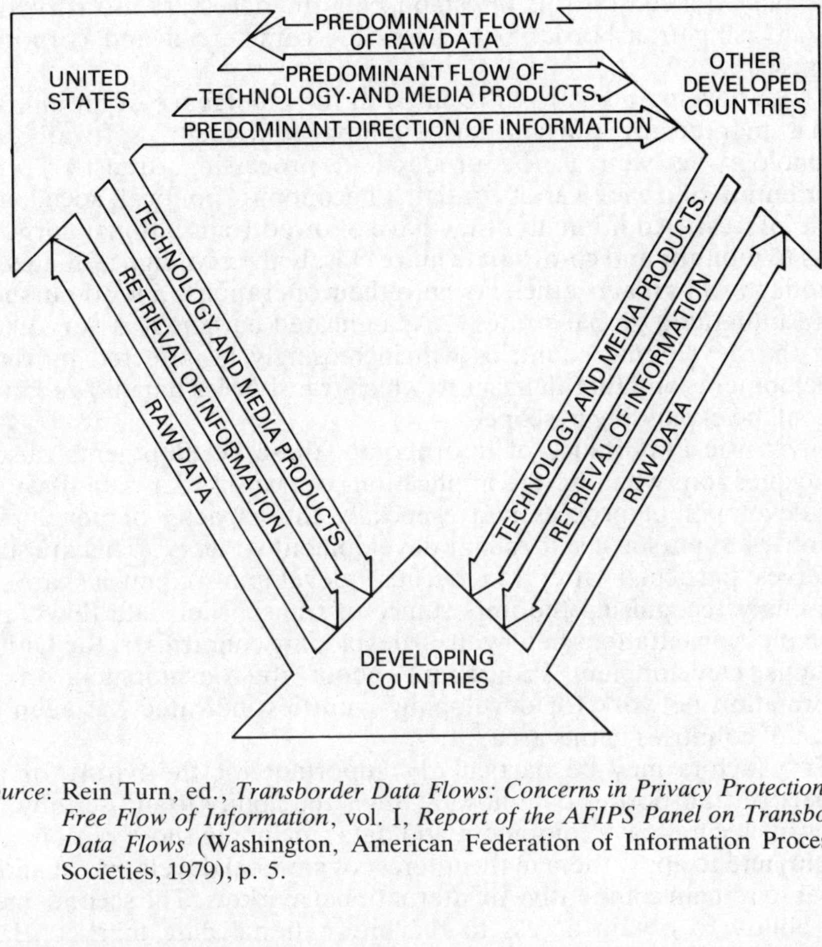


*Source*: Rein Turn, ed., *Transborder Data Flows: Concerns in Privacy Protection and Free Flow of Information*, vol. I, *Report of the AFIPS Panel on Transborder Data Flows* (Washington, American Federation of Information Processing Societies, 1979), p. 5.

## Implications for host countries, particularly developing ones

### *General considerations*

Information is a resource. It makes alternatives known, reduces uncertainty about their implications, and facilitates their implementation. Hence, information, appropriately applied in the pursuit of well-defined objectives, is central to decision-making. And, increasingly, decision-making depends on access to a great variety of data, which has to be stored and processed on computers to make them manageable.

Telematics has given a new importance to information because it has made possible unprecedented flows of data. It has also led to the establishment of transnational computer-communication systems and hence given rise to transborder data flows. Transnational corporations have played a central role in establishing the necessary infrastructure and in using transborder data flows for commercial and corporate purposes.

These developments have resulted in the emergence of an international market for transborder-data-flow equipment, software and technology, as well as the production, processing, transport, and distribution of a wide variety of data of economic, political, social, and legal interest. Additionally, they have allowed transnational corporations to monitor and co-ordinate more closely their foreign affiliates, to introduce still greater efficiency into their operations, and to pursue a more integrated global strategy. As indicated earlier, it is very likely that the developing countries will increasingly be affected by these developments and that the uses to which transborder data flows can be put will be expanding in scope.

Given the importance of information, these developments raise a basic question: what are the implications of transborder data flows for the development process and especially the capacity of developing countries to pursue a self-reliant development strategy? This question deserves particular attention because developing countries are increasingly recognising the importance of transborder data flows. For example, consultations held with 30 developing countries by the United Nations Development Programme about the development of an information network for developing countries indicated the keen interest of countries in this area.[8]

Two factors may be particularly important for the nature of the impact of transborder data flows. One is the ability to utilise fully the relevant technology, equipment and data (including those of domestic origin) and to apply them in the interest of national development and in order to remain competitive in international markets. The second one is the ability to obtain access to the international data market. Both factors, probably as mitigated by the level of development, influence the participation of the developing countries in transborder data flows.

The impact of transborder data flows may even reach beyond the development process. Since information is central to decision-making concerning most political, economic, social, and legal matters – domestically as well as internationally – transborder data flows are also likely to affect the general relations between developed and developing countries. More specifically, to the extent that information is a basis of power, access to information and the ability to utilise it can give countries and institutions (e.g. transnational corporations) political, economic, social, and legal advantages over other countries and institutions (e.g. trade unions, domestic corporations).

Since transborder data flows are a new phenomenon, their implications have neither manifested themselves clearly nor have they been the subject of much empirical research. By necessity, therefore, the following discussion of the impact of transborder data flows can only be exploratory and tentative. It endeavours to raise issues, discuss possible implications, and point to potential costs and benefits. The discussion gives more attention to the potential costs and conflicts associated with transborder data flows, because these could prevent the realisation of the full benefits of this new phenomenon.

## *Potential benefits and problems associated with transborder data flows*

*Potential benefits* As suggested earlier, the implications of transborder data flows depend on the user's ability to obtain access to the international data market and to utilise fully the underlying technology and equipment. If both conditions are fulfilled to a reasonable degree and if transborder data flows are applied in ways that are useful for developing countries, the benefits of these flows – immediate and future – could be considerable.

Most importantly, transborder data flows could give a growing number of users instant access to a rapidly expanding and diversifying pool of up-to-date knowledge. Such access, in particular to data that are of direct interest to developing countries, may give developing countries, in certain sectors, more information on alternatives and better access to data needed for implementing the alternatives selected. This may be of interest, for instance, for negotiations with transnational corporations, since an increasing amount of business and economic data is becoming available in machine-readable form. Furthermore, remote-sensing data could put countries in a better position to monitor their own resources, to assess their value, and to determine factors affecting their optimal use.

Equally important, if national corporations in developing countries avail themselves of some of the technology that is being applied by transnational corporations from developed market economies, they may maintain or increase their competitiveness in world markets.

Finally, to the extent that transborder data flows are a tool for stimulating scientific and technological progress and for facilitating the transfer of a whole range of data required for accelerating the development process, they can contribute to a more efficient international allocation of resources and thus constitute a source of increased productivity and economic growth. In fact, considering that the international data market contains substantial information about developing countries, transborder data flows could become a helpful tool for economic co-operation among developing countries. The objective of economic co-operation among developing countries could be particularly furthered if networks were to be constructed which, using appropriate technology and supplying appropriate information, link developing countries among themselves.

*Potential problems* Apart from financial costs,[9] the potential costs associated with transborder data flows stem also from some of the above-mentioned features of the present structure of the international data market and the nature of the new technologies. Most notably, these costs relate to the international division of labour, the ability to establish national telematics capacities, the negotiating capabilities of developing countries, national sovereignty, and vulnerability.

The international division of labour associated with the structure of the international data market, as well as some of the grievances held by developing countries against it, is familiar from other areas of relations between developing and developed countries, e.g. in that of natural resources. For many developing countries, a key issue is that most of the value-added direct benefits resulting from the over-all production process occur at the processing and distribution stages and hence do not normally accrue to developing countries. Similarly, certain spin-off effects (such as incentives to research and development, the stimulation of related economic activities and, in the specific context of transborder data flows, the production of software) also may tend to be associated with processing and distribution rather than with the production of the raw material – be it mineral resources or data. From this perspective it has been observed that data export for processing abroad comes to be regarded in an information age in the way the export of raw materials for manufacturing abroad is regarded in an industrial age – as the best way to fall behind; as a mark of underdevelopment'.[10]

Yet, the international division of labour is not a static phenomenon. In fact, the very technology that underlies transborder data flows may enable the developing countries to modify the present structures. At present, however, the prevailing division of labour makes it difficult for developing countries to establish national telematics facilities and may not leave much room for creating self-reliant capacities. Furthermore, this situation tends to be self-perpetuating. If data must be exported for processing, existing or potential national service bureaux are deprived of prospective customers. As a result, domestic demand for hardware and software may not grow as rapidly as it otherwise might have done. Thus, some countries are confronted with a difficult choice of balancing the immediate benefits associated with the import of data and data-processing services against the sacrifices associated with the long-term benefits that could result from a national developmental policy.

The now prevailing pattern of corporate transborder data flows suggests that transnational corporations cannot be expected to be of decisive help to host countries in solving this problem. First, transnational corporations appear to be major exporters of data: in the case of Canada, for instance, approximately 90 per cent of total net transborder data flows involve data flows from foreign affiliates in Canada to parent corporations abroad. Second, the transnational intra-firm division of tasks within corporate computer-communication systems often allocates input-output or very simple processing functions to facilities located in developing countries, while the more powerful equipment and its production tend to be concentrated in the developed countries. Finally, the information advantage of transnational corporations may be an additional advantage to them, placing domestic enterprises at a competitive disadvantage and thus hindering the emergence of indigenous capacities in host countries. The last factor also bears directly on the bargaining strength of these corporations *vis-à-vis* states and groups within states (e.g. trade unions).

Uneven access to the international data market and varying capabilities to utilise fully the underlying technology and equipment may even affect, to the extent that transborder data flows involve the 'migration of key decision-making functions'[11] to foreign locations, a country's ability to influence the direction of its political, economic and cultural changes. In addition, national sovereignty may be impaired if knowledge about the full range of alternatives open to a given country in a given situation is restricted because of lack of access to relevant information or because of an underdeveloped capacity to apply the relevant technologies. If a country lacks data about itself and the international system in which it finds itself – because of a very limited capacity to collect, access, or process them – it can be said to lack pertinent decision-making capacity about its own future. Seen from this angle, an extension of the concept of sovereignty to include 'informational sovereignty'[12] is not surprising.

A possible erosion of such informational sovereignty would connote an increase in vulnerability – an issue explicitly taken up by a study of the Government of Sweden.[13] For example, a disruption of critical data imports (e.g. because of a breakdown of computer systems, strikes, natural disasters, acts of terrorism, or political pressure) or the export of sensitive data for purposes of processing (e.g. name-linked data, in particular when exported to data havens, i.e. countries with lax or no data-protection laws), could have palpable effects on a country and its citizens and could even expose them to foreign manipulation. This vulnerability may be particularly high in the cultural sector if technological developments in the communications industries (especially the mass media) should allow point-to-point flows of messages and the associated values and behavioural patterns. The impact of such a development could be particularly felt because the international communications market is serviced by a relatively small number of transnational corporations based in a few developed market economies.

## *Prospects*

The present situation poses a dilemma for developing countries. If they make vigorous attempts to be integrated into the international data market, many of them may face the prospects of increased dependence on imported technology and equipment; on the other hand, if they stand aloof, they may have the problems associated with a very limited access to the rapidly expanding pool of machine-readable data.

The solution may lie in selectivity. A possible point of departure could be the identification of those data that are of particular importance to developing countries in that they meet their specific conditions, needs, and objectives. These, in turn, could be matched with technologies that are geared to these purposes, that have acceptable secondary impacts (e.g. in terms of the domestic distribution of benefits), and that allow applications which can be supported by existing or feasible infrastructural arrangements and skill endowments. Depending then on the over-all development priorities, full access to, and reliance on, the international data market may be necessary to fulfil some of these priorities; for others, however, it may be more appropriate to examine other alternatives or to build telematics capacities and networks that are directly geared to the development priorities of the developing countries. Particular attention would have to be paid to a careful weighting of the (financial and social) costs and benefits of transborder data flows.

It has to be re-emphasised, however, that the novelty of the phenomenon and the scarcity of pertinent empirical research do not permit, at present, firm conclusions and forecasts about the impact of transborder data flows. But it seems safe to say that the potential benefits of transborder data flows are considerable and the potential problems related to them are important. It also appears that conflicts are more likely to arise if potential problems are not identified and understood in time and if equitable solutions are not actively sought.

## Guidance frameworks

The increasing awareness of the importance of information and the growing realisation of the possible political, economic, social, and cultural effects of transborder data flows have led national Governments to search for adequate responses to this new phenomenon. Reflecting the longer and more intense experience of the developed market economies with telematics and transborder data flows, these countries have taken the lead in the process. So far, however, the search for suitable guidance frameworks at the domestic and international levels has only commenced – with the notable exception of questions pertaining to transborder flows of personal data. Concerns about safeguarding the privacy of these flows, in particular, have served to focus attention on transborder data flows in general, triggering national and, eventually, international actions. In fact, these actions have resulted in the first international legal frameworks designed specifically for transborder data flows.

### *Personal transborder data flows*

Computerised personal data systems may have adverse effects on the right to privacy because, *inter alia*, the amount and range of data stored on the private life of an individual faces virtually no limitation. A number of countries have, therefore, enacted privacy-protection laws in this field.

Given the transnationalisation of computer-communication systems, compliance with these laws can only be ensured if the same data-protection standards are recognised in all countries involved or if the flow of personal data from one jurisdiction to another is subject to a common regime. It is in this context that transborder flows of personal data have been made subject to national and international rules. Concerning the latter, both the Council of Europe and the Organisation for Economic Co-operation and Development have adopted instruments specifically addressing the subject of transborder flows of personal data.[14]

### *Non-personal transborder data flows*

For many countries, however, the importance and impact of the new technologies go beyond the realm of privacy and touch upon the very nerve of their economic development and international relations. A growing number of countries have therefore taken steps to deal with the non-personal data aspects of transborder data flows. Yet these steps focus mostly on telematics (i.e. are not specifically addressed to the transborder-data-flow dimension of the problem) or are limited to the formulation and adoption of policies (i.e. do not have the nature of laws and regulations).

A number of countries or groups of countries, however, have begun to design policies for dealing specifically with non-personal transborder data flows. Most notable among them are Canada, the European Community, France, Sweden, and, in particular, Brazil.

The Brazilian effort to deal with transborder flows of non-personal data grew out of that country's national computer policy – a policy that aims at facilitating the creation of national capabilities – and it continues to be an integral part of it. Since 1972, the Coordinating Commission on Data Processing Activities (CAPRE), a part of the Planning Ministry, has supervised the use and acquisition of computers, first only those of the Federal Government of Brazil and since 1976 also any other computer or computer part. In the framework of its industrial policy in this area, action was eventually taken to deal with transborder data flows.[15]

According to legislation enacted in May 1978, transnational computer-communication systems became subject to approval by CAPRE. In particular, the regulation requires applicants for such systems to fill out a detailed form in which they have to provide information about themselves, other parties participating in the proposed service, the reasons for the application, and the resources involved. The Executive Secretariat of CAPRE analyses the applications in terms of their micro- and macro-economic impact and their impact on privacy and sovereignty. The Council of CAPRE subsequently decides whether or not an application should be approved. Approval is granted for a maximum of three years; conditions may be specified. Between May 1978 and January 1980, 19 applications were filed and decisions were taken on 16. Approval was denied for applications related to the use of time-sharing services and data banks abroad and certain types of the international operations of foreign affiliates; approval was given for airline reservation systems and demonstration systems. In general, the Government of Brazil 'does not allow the use of computers placed abroad, which through teleinformatics accomplish tasks whose solution can be obtained in the country'.[16]

The experiences gained from the application of this regulation were taken into consideration when the Brazilian rules governing data transmission were finalised and published at the beginning of 1979. At the same time, the Special Secretariat of Informatics was established (as part of the National Security Council) as a direct advisory body to the President of Brazil. This body, which succeeded CAPRE with a broader and stronger mandate, is responsible for the formulation and implementation of the National Informatics Plan, including all questions related to the manufacture and use of computers and electronic components and all questions pertaining to transborder data flows. These actions reflect the great importance which Brazil attributes to telematics and transborder data flows. It is an integral part of an approach that recognises that 'the informatisation of human activities will be decisive in determining a new international economic order, where the roles of the different countries will be dictated by the care with which they have established their respective policies of transborder data flow' and that, therefore, 'in terms of strategy, it is fundamental that a country have control over the information resources essential to its sovereignty and development'.[17]

Other countries are guided by different considerations. A number of developed countries, most notably the United States, insist that any framework for non-personal transborder data flows must be guided by the recognition of the principles relating to a free flow of information, free enterprise, free trade, and the adequate availability of appropriate telecommunications facilities and services.[18]

The policies pursued by Brazil and other countries indicate an interest of some governments to take measures aimed at averting various perceived undesirable political, economic, social, and cultural effects of transborder data flows. The scope of such measures is considerable and all industries and entities engaged in transborder data flows could be affected by them. The result may well be a proliferation of a variety of national measures regarding non-personal transborder data flows which could obstruct the realisation of the positive effects of these flows.

The prospects of divergent developments at the national level are compounded by the fact that an international framework directly pertaining to the transborder flow of non-personal data as a phenomenon *sui generis* does not presently exist. One major effort has, however, recently commenced that could conceivably lead to the establishment of such a framework: an *ad hoc* Expert Group on Transborder Data Flows and Data Communications Policies of the Organisation for Economic Co-operation and Development has the mandate to investigate the legal and economic problems related to the transborder flow of non-personal data with a view of developing guidelines in this area as well. A programme of research on these issues is already underway.

The developing and socialist countries are, however, so far not known to have initiated collective discussions on the adoption of a common stand specifically designed to deal with the impact of transborder data flows and the role of transnational corporations in them.

## Possible areas of future work

At present, there exists no 'legal infrastructure to sustain the transition to the information economy'[19] nor, one may add, the transition to a world economy in which commercial and corporate transborder data flows of transnational corporations are bound to play a growing role. This lacuna exists although it is increasingly recognised that information is a key ingredient in decision-making, especially regarding the economic development of both developed and developing countries.

This situation may represent both an opportunity and a challenge. In a relatively undefined legal environment, and with respect to an activity that is only beginning to become important, vested interests are not omnipotent and positions are not frozen. Hence, preconditions for a co-operative approach that balances the legitimate claims for national sovereignty and development with the very real need for transborder data flows are still favourable. The challenge is to seize this opportunity and to consider how it can be ensured – through the strengthening of national capacities and, if the need arises, through collective action – that the entire international community can participate in the benefits associated with telematics and transborder data flows. For the developing countries, in particular, the ability to access the international data market and to utilise fully the underlying technology and equipment may, in fact, be an important element for the realisation of the new international economic order. Transborder data flows could lead to an accentuation of existing imbalances between the developed and the developing countries, but they can surely also be channelled towards redressing these imbalances.

Policy-formulation regarding transborder data flows requires a thorough understanding of the implications associated with this phenomenon. In view of the paucity of empirical research findings it is very difficult to offer precise conclusions about the multifaceted effects of these flows. This is particularly unfortunate since many developing countries might, on the one hand, want to speed up the introduction of transborder-data-flow technology in order to reap its potential benefits, while, on the other hand, they will want to avoid the potential problems associated with it. Thus, a full appreciation of the possible costs and benefits of transborder data flows and an examination of the conditions under which the application of these flows can be most useful are required.

Obviously, this points to the importance of research. The possible research agenda is broad. Since the work of the Organisation for Economic Co-operation and Development focuses largely on concerns of industrial states, other research could now assure that sufficient attention is paid to questions of primary interest to developing countries. Since transnational corporations are of key importance in practically all aspects of transborder data flows, research would naturally focus on their role and contribution.

Probably the most important question to examine empirically and in detail would be the impact of transborder data flows on developing countries and the role of transnational corporations in it. In the same manner in which a number of developed countries – notably Canada and France – have found it necessary to prepare in-depth evaluations of the effect of transborder data flows on their societies and economies, it may also be useful for individual developing countries to study the consequences of this new phenomenon for their societies and economies, particularly in light of their development objectives. After a number of such case studies have been conducted, some generalisations could conceivably be made that would permit conclusions about the type of actions that may be advisable. At the same time, it may be possible to indicate to what extent the experiences of developed countries in this area are relevant and useful to the developing countries.

Such work would have to identify the possible areas of application of transborder data flows in the light of over-all development objectives, i.e. the specific purposes they are meant to serve. It would also have to aim at establishing an inventory of information resources that can be mobilised individually or collectively by developing countries to fulfil these purposes. In this context, it would furthermore be useful to analyse how the objectives and identified resources could be translated into national production and distribution structures and how the performance of these structures could be evaluated and monitored. Moreover, it would, of course, also be interesting to study which resources could be more advantageously obtained from the international market and through transnational corporations.

Another topic of research could relate to the question of access to the international data market in which transnational corporations play a central role. On the basis of a thorough analysis of the dimensions, nature, and structure of this market and especially the type and costs of services and data available, alternatives could be determined as to how access to this market could be improved and how its resources, in a complementary manner, could best be utilised for the objectives of the developing countries.

A third area of possible research is closely related to the preceding two, and concerns the negotiating capacity of host, especially developing, countries. A key question here is how transborder data flows could best be used by host countries to assist them in negotiating advantageous contracts and agreements on the whole range of their interactions with developed countries in general and transnational corporations in particular.

Since a number of countries have initiated policies addressed to transborder data flows, of which most affect directly the commercial and corporate operations of transnational corporations, it may also be useful to examine the experiences of these countries in formulating and implementing their policies, particularly in light of the objectives they have set themselves.

Finally, more needs to be known about the effects of the application of transborder data flows on the relations between host, especially developing, countries and transnational corporations. Questions of particular interest in this context concern the impact of transborder data flows on the centralisation or decentralisation of key corporate functions, the intra-firm division of labour, training of the labour force, industrial relations, and the ability of foreign affiliates to pursue a policy that is responsive to the needs of host countries.

Not all these possible areas of research can be tackled at the same time. But it may be desirable if work began without delay, lest developments reduce the range of options that may currently still be open. Furthermore, the awareness of the implications and potential of transborder data flows could usefully be raised through documentation, seminars and other means in order to contribute to a demystification of this new phenomenon. The principal objective of such efforts would be to devise national mechanisms through which a maximum of benefits can be obtained from the application of transborder data flows, while minimising their potential negative effects.

## Notes

1 Organisation for Economic Co-Operation and Development, *INTERFUTURES: Facing the Future. Mastering the Probable and Managing the Unpredictable* (Paris, 1979), pp. 114 and 336, respectively.

2 The term 'telematics' (or télématique, which is used in this report) can be used interchangeably with 'compunications' and 'teleinformatics'.

3 Data are symbols that can be handled and transmitted through computers; information is the combination of data into messages intelligible to human beings – that is, it is the processed form of the raw material, data.

4 See, in this context, United Nations Educational Scientific and Cultural Organisation, *Many Voices, One World: Towards a New More Just and More Efficient World Information and Communications Order*. Report by the International Commission for the Study of Communication Problems (Paris, 1980).

5 Raymond Vernon, *Storm Over the Multinationals: The Real Issues* (Cambridge: Harvard University Press, 1977), p. 2.

6 See 'New horizons for international co-operation' (DSTI/ICCP/80.20), a BIAC position paper presented at the High-level Conference on Information, Computer and Communications Policies for the 1980s (6–8 October 1980) p. 3.

7 D. A. Woodland and P. S. Doepel, 'Management of Distributed Organization,' in E. J. Boutmy and A. Danthine, eds, *Teleformatics '79* (Amsterdam, North Holland, 1979), p. 10. It should be noted that the processes described here refer to a decentralisation of equipment usage and not necessarily of control.

8 See the project description paper entitled 'Development information network' being prepared by the United Nations Development Programme.

9 Principal financial costs consist of establishing the necessary computer and telecommunication infrastructure, leasing the lines required for a network, subscription fees to user-group networks, and data-base retrieval costs. In the case of SWIFT, for example, an initial entry fee of about $50,000 is required, together with connection costs of about $5,000, the price of an interfacing device of about $70,000–$75,000 and a per-message fee for the actual usage of the system. The costs of data-base retrieval consists of about $1,500 for an appropriate terminal, on-line retrieval expenses and off-line printing charges. In most cases, on-line retrieval is charged on the basis of on-line connect time, with hourly rates normally ranging from $15 to $90. Printing charges per full record of information, in most cases, does not amount to more than $0.50. In countries served by TYMNET or TELENET, communication costs are based on distance-insensitive per-digit charges. In all other cases they amount to the costs of a direct international telephone call.

10 Keynote address delivered by Charles M. Dalfen, Vice-Chairman of the Canadian Radio-Television and Telecommunications Organisation at the opening session of the IBI World Conference on Transborder Data Flow Policies. Reprinted in *Transnational Data Report*, 3 (July/August 1980), p. 6.

11 Peter Robinson, 'Strategic issues related to transborder data flow', *Telecommunications*, 13 (May 1979), p. 85.
12 Allan Gotlieb, Charles Dalfen and Kenneth Katz, 'The transborder transfer of information by communications and computer systems: issues and approaches to guiding principles', *American Journal of International Law*, 68 (1974), p. 236.
13 See Government of Sweden, Ministry of Defence, *The Vulnerability of the Computerised Society* (Stockholm, 1978).
14 See, respectively, the Convention for the Protection of Individuals with Regard to Automatic Processing of Personal Data, adopted by the Council of Europe in 1981 (*European Treaty Series*, No. 108) and the Recommendation of the Council Concerning Guidelines Governing the Protection of Privacy and Transborder Flows of Personal Data adopted by the Council of the Organisation for Economic Co-operation and Development at its 523rd meeting on 23 September 1980), (C(80) 58 (Final) ).
15 See Mario Dias Ripper and Jorge Luiz Cesário Wanderley, 'The Brazilian computer and communications regulatory environment and transborder data flow policy', a paper presented at the IBI World Conference on Transborder Data Flow Policies, Rome, 23–27 June 1980.
16 Keynote address delivered by Joubert de Oliveira Brízida, Chairman of the Brazilian delegation, at the opening session of the IBI World Conference on Transborder Data Flow Policies. Reprinted in *Transnational Data Report*, 3 (July/August 1980), p. 33.
17 Ibid. p. 34.
18 See, for instance, the keynote address by the Honourable Matthew Nimetz, Under-Secretary of State for Security Assistance, Science and Technology, Department of State, Government of the United States, at the High-level Conference of the Organisation for Economic Co-operation and Development on Information, Computer and Communications Policies for the 1980s, Paris, France, 6 October 1980. Mimeographed document.
19 See the opening address by Mr André Giraud, Minister of Industry, Government of France, at the High-level Conference of the Organisation for Economic Co-operation and Development on Information, Computer and Communications Policies in the 1980s. Mimeographed document, p. 7.

# 4

# Information Access to International Finance: What Problems for Developing Countries?*

Stephany Griffith-Jones

## Introduction

In this chapter some issues of particular relevance to improved information access for developing countries in the field of international finance are explored. The first part deals with business negotiations between particular developing countries' governments or firms with international financial institutions, mainly the multinational banks. The second deals with international negotiations concerning changes or reforms of the world financial system, in which developing countries participate or whose results will have an important effect on their economies. This section will deal not only with improvement of information for negotiations in which developing countries participate (for example, UN Global Negotiations. Mexico Summit), but also information about and possibly access to the very crucial negotiations and decisions made in discussions where developing countries do not participate, (including Group of 10, and its institutions like the BIS).

## Information and risk assessment: the banks

There are considerable problems of information gathering and assessment by developing countries in relation to types and conditions of available financial flows. Before concentrating on international capital markets, it seems important to emphasise that insufficient information and inadequate assessment is not only a 'one way' problem; it operates in both directions.

* I am very grateful to Carlos Diaz-Alejandro and Rita Cruise O'Brien for useful comments on an earlier draft.

Although multinational banks are often reluctant to admit it publicly, the information which they have about their developing countries' clients has several very important limitations. Most multinational banks have limited knowledge about smaller economies and/or about those countries which borrow less frequently. Smaller banks have, in general, very little information about Third World countries, as they have very reduced staff who work in this area. Smaller banks rely directly on signals about whether larger banks are lending to a particular country, or whether it has signed a loan agreement with the IMF, rather than on their own information and analysis. This type of behaviour was clearly described by Henry Wallich, Member of the Board of Governors of the Federal Reserve System,[1] when he said, '. . . There are other cases where deterioration was progressive and the eventual failure of debt service should not have been very surprising, as for instance, in Zaire and Turkey. *The only explanation why some banks have wanted to make a loan under these circumstances would seem to be that other banks were making loans*' (author's italics).

Smaller banks also rely greatly on published reports and information produced by the largest multinational banks, the IMF and the World Bank. If they can get access, then they also use confidential, unpublished information produced by these institutions; on the whole, small non-US banks have greater difficulty in obtaining access to the long confidential reports produced by the IMF and the World Bank than large US banks which tend to have better personal contacts in those institutions.[2]

Furthermore, it is only some of the larger banks which have detailed knowledge of the IMF policies in different credit tranches; these more sophisticated banks know more about details of Fund policies in the credit tranches.[3] For example, they can distinguish clearly a first credit tranche programme between a country and the Fund, with its lack of performance clauses, from the higher tranche credit arrangements with phased drawings and performance clauses. Such banks have on occasions delayed their lending until the country has a credit arrangement in the higher tranches with the IMF and have related their own lending to continued observance of the performance clauses in that programme; in practice, this has implied parallel operations between those private banks and the Fund, in which the former rely to a great extent on the latter's judgement about the 'appropriateness' of the country's adjustment policies for improving that country's 'creditworthiness'.

Large banks, particularly those heavily involved in international lending, also rely extensively on their own 'economic intelligence' departments, as well as on private consultancy firms which carry out specialised 'political risk assessment'. From the mid-seventies these departments and consultancy firms grew very rapidly. Their main task has been to evaluate developing (and other category) countries' 'creditworthiness', that is a country's *future* ability and willingness to service and repay its loans. The methods of assessment vary widely, with the big American banks usually relying to a relatively large extent on 'sophisticated' methods, including elaborate computer-based mathematical models. It is however unclear whether the methods used for risk assessment are the most suitable ones. More fundamentally, it seems increasingly doubtful as to whether it is possible at all to design an adequate methodology to project future political or even economic developments. The political developments in Iran, Poland and, more recently, in Argentina (as well as their effects on those countries' ability to service and repay their debts) were clearly not forecast with accuracy by the leading private banks, which had been regarding those countries as quite 'creditworthy'. It seems that it is only with difficulty that such political events could have been forecast accurately.

The record of economic forecasting has also been particularly unimpressive in recent years. The international financial institutions involved include not only the large multinational banks but also the IMF and the World Bank, which have highly qualified professional staff, substantial data banks and sophisticated mathematical models. All of these have tended to misjudge rather seriously the projections for the real price of oil for the early Eighties. Naturally, wrong forecasts of the price of oil affect economic projections for most developing countries rather significantly.

It can be concluded that private multinational banks also have serious problems as regards information crucial to them about their developing country clients. Such problems are particularly serious for the smaller banks, with little or no staff involved in obtaining or evaluating in some detail both published and unpublished information. It seems that imperfect information has contributed, particularly, but not only, in the case of smaller banks, to 'fads' or 'herd behaviour' in private bank lending.[4]

## Information problems: developing country borrowers

Developing country problems of information access to international finance are very often linked to broader issues of macro-economic management. As individual operations on the Euro-markets are so large (rarely falling below US$ 50 million), the issue of whether and on what terms a particular loan should be raised can often be linked to the broader question of whether the country should be borrowing at all at the time. The latter decision depends also on many non-financial and even non-economic considerations, such as the priority given to autonomy from external finance.

In this respect, it is important to evaluate the need and costs of private specialised information and advisory services offered by certain investment and merchant banks to developing countries. These banks offer 'objective advice' in such varied areas as arrangement and negotiation of conditions on new loans, renegotiation of existing loans, investment of foreign currency reserves and even organisation of the domestic monetary systems. The emergence of banks specialising in advising developing countries, include amongst the best known, but not only the 'triad' of Kuhn Loeb Lehman Brothers International in New York, Maison Lazard et Cie in Paris and S. G. Warburg & Co. in London. They responded largely to the situation in which investment banks, which often managed the lending syndicates in the early Seventies, are now unable to maintain that role, which is almost being monopolised by the large commercial banks, who have the vast financial resources required for the actual lending. As a result, these merchant banks have found a profitable form of participating in the lending process by acting as consultant on or for developing countries.[5]

It seems an important area of study (both for academics and for policy-makers) to determine whether, for particular developing countries, the benefits obtained through such consultancies, including additional inside information or contacts, outweigh the costs. These costs are both financial, considering the magnitude of the fees charged, and non-financial, given the fact that a report by an influential bank or group of bankers may influence a country to pursue policies inconsistent with national economic and political targets. In evaluating the benefits, it seems necessary to explore whether the additional information is necessary at all, and whether it cannot be more cheaply obtained through other mechanisms. Amongst the alternative options are to provide high-level training for nationals in this area, the purchase of technical advice from other more experienced developing countries, or through a pool of experts used by a group of developing countries. The option or options chosen should naturally have to ensure both lower financial cost and greater consistency with the country's national targets than the consultancies carried out currently by merchant banks.

In order to consider these and other policy options for obtaining information, different categories of developing countries must be distinguished. Frequent and infrequent borrowers have quite different information problems. Countries also differ according to their level of development and bargaining capacity, cumulative experience in international finance and the degree of sophistication of domestic financial institutions and expertise.

In this respect, it would seem a plausible hypothesis to assume that some countries (for example, Brazil, Mexico, South Korea) which are frequent borrowers on the Euro-markets, have a relatively high level of development and large cumulative experience in this field, and suffer very few information problems in this area. It must be noted, however, that information problems may exist, similar to those that could arise also for a developed country approaching the market. Other countries, particularly those which are smaller, more remote geographically and/or in terms of communication, and operating on the Euro-markets infrequently, have much more substantial information problems. Frequency of operations, as well as proximity to the markets in informational terms is particularly important in the case of international finance because the markets' conditions and even mechanisms change very rapidly and frequently. Even though some information on changing conditions, including spreads, is published with certain regularity, other information such as levels of management and front-end fees is much harder to obtain from published sources. Informal sources of information are therefore particularly valuable.

## The need for Third World cooperation

An important conclusion may be established concerning the need to expand Third World cooperation for information sharing in this area. As pointed out above, this can be carried out both at a multilateral level, via the UN or regional institutions, and on a bilateral basis. A possible problem could perhaps be whether countries which have important experience in the Euro-markets would be unwilling to reveal the 'tricks of the trade' to possible competitors in the markets.

There are two important sources of finance to developing countries about which information is probably even more limited than that described above. There are, for example, financial sources in the industrialised countries which have been barely tapped by developing countries like pension funds and insurance companies' funds. Although there are certain structural and institutional barriers, information problems on both sides seem to be playing an important role in restricting potential access.

Recently there has emerged a large number of rapidly growing Third World financial institutions about which information is relatively limited. Not even The Group of 77 has at present a list of those new Third World institutions, their size, types and conditions of their loans, areas they cover, etc. Individual countries can only gradually and with difficulty gain access to this information. A very useful basic task for improving developing countries' information about international finance would be to compile a sort of Who's Who on the Third World financial institutions, which should include both private and public financial institutions.

Awareness of the vast and varied financial institutions of the Third World would not only facilitate greater access to their finances by developing countries, but could also lead to the development of more intra-developing countries technical links. These could replace or complement those with Northern financial institutions.

## Information access and international monetary reform

One of the most crucial areas of information on international finance relates to the broader negotiations on changes and reforms of the world financial system. In this field, there is a near monopoly of *detailed* knowledge of the international financial system by rich country experts, thus limited the possibility of suggesting changes in the system which would be in the interests of developing countries.[6]

Furthermore, the rich countries have special institutions where they discuss first (and sometimes *only* amongst themselves) important changes or reforms of the international financial system, which have an important impact on Third World economies. Thus, many of the crucial reforms are not even discussed in institutions such as the IMF (where Third World countries participate, even though they have a minority in the vote) but in *rich country institutions*, such as the Bank for International Settlements which services the central banks of industrialised countries. For example, the crucial issue of whether, how and when to control the Euro-markets has been discussed almost exclusively within the framework of the Bank for International Settlements, even though it will have very important effects on developing countries, amongst the largest depositers and lenders on these markets. This is only one important example of many such issues discussed only or mainly in those institutions. The developing countries ought now to demand increasing access to those negotiations, not only because it would allow them to participate in crucial decision-making, but also because it would allow access to very important inside information.

## The need for a secretariat for the Third World

The developing countries must not only increase their access to all important fora where decisions on international finance are made, but also improve their participation in such negotiations. An important element in achieving this would be the establishment of a research and information centre designed to serve the needs of these countries and linked closely to these negotiations. This would to a certain extent provide a necessary counterpart in the South to such Northern institutions as OECD, BIS and others.

It is unfortunate that at the 1981 Group of 77 meeting held in Caracas, as well as at former meetings, agreement could not be reached on a G77 Secretariat, nor even on the more limited objective of a G77 research and information centre.[7] The main problems for setting up such a Secretariat or Centre seem to be political. Some countries (the larger and richer ones) do not feel they require a Third World Secretariat, as they have confidence in the expertise of their own diplomats. The real reason, perhaps is that they fear such a Secretariat would imply loss of their relative predominance. Other countries would prefer regional institutions within the Third World (i.e. Latin American, OPEC, Asian). Still other countries have ideological reasons for resisting a Third World Secretariat or information centre.

Given the political constraints, it seems desirable that proposals are put forward which will achieve some of the targets of a G77 Secretariat by more modest means, including regional secretariats, technical groups on specific issues, for example financial. The gains to the Third World of improved information in negotiations may hopefully in the medium-term begin to outweight the political obstacles, leading to a G77 Secretariat with an information component.

Another path which may lead to improved information flows for developing countries may be through the development of Third World financial institutions, by the continued growth of existing ones and/or the creation of new ones, such as a Third World Development Agency. Much of the rich-country expertise (particularly on international financial issues) is concentrated in and around the World Bank and the IMF. By obtaining greater influence within these Bretton Woods institutions, Third World countries might also improve access to important information sources, as well as their access to decision-making. Furthermore, the development of parallel or complementary institutions to the Bank and the Fund, with greater Third World control would inevitably lead to an enrichment of much-needed expertise on international finance in the service of the South.

## Notes

1 '*LDC Debt – To Worry or Not to Worry*', speech at the 59th Annual Meeting of the Bankers' Association for Foreign Trade, Boca Raton, Florida, June 2, 1981.
2 It is interesting to note that the IMF staff itself was concerned about inequity of information access amongst different categories of private banks and even expressed this concern in an internal document entitled *Fund Relations with Commercial Banks*, June 1977.
3 Speech of Henry Wallich, op. cit. and interview material.
4 See G. K. Helleiner on the problem of 'fads'.
5 For a brief, but candid, account of investment/merchant bankers' role in

'advising' developing countries, see 'Gamekeeper turns poacher', *Financial Times*, July 17, 1979.

6 This point was first drawn to my attention by Carlos Diaz Alejandro's paper, 'International finance: a subject of special interest for developing countries', Seminar on International Economics and Developing Countries, Priorities for Research, organised by CIEPLAN and the Ford Foundation, Santiago, Chile, May 1978.

7 At a meeting of senior representatives from 44 developing countries held in New Delhi in February 1982, the proposal presented by Julius Nyerere of a technical secretariat support group for determining areas of possible negotiations with the North was similarly rejected. *Source: South*, May 1982, Editorial, p. 4.

# 5

# The Role of Information in Commodity Negotiations: The Case of Coffee

Roberto Junguito and Diego Pizano

The object of this chapter is to explore the importance of information access to the outcome of the bargaining processes in the world coffee market. It starts with the general issue of information in economic models, followed by a specific section on the structure of the world coffee market. The main part of the chapter considers information variables and bargaining power within the framework of international coffee agreements and producers' marketing arrangements.

## Conceptual background

In economic theory there have been several different approaches to the information factor. General equilibrium theorists, for example, have explored the conditions required for convergence toward stable equilibrium positions. As Arrow and Hahn[1] have shown, it is a necessary condition to assume that uncertainty in different economic agents must gradually disappear if an equilibrium position is to be reached. If there is no perfect foresight on future conditions it is assumed that the agents develop a set of rational expectations which ensure the transition towards intertemporal equilibrium. Such authors have traditionally ignored concepts such as strategies, reciprocal expectations and uncertainty.

Such issues are, however, recognised and considered analytically in game theory. In their classical description of a game Von Neuman and Morgenstern[2] seek to show that if a move made by a player corresponds to a strategy (contrasting with a random move), then the information held by the player becomes a crucial variable in the problem. The authors were interested to establish the conditions which generate pre-determined solutions to different games. They showed that in situations of incomplete information there is no determinate solution to the game.

Von Neuman and Morgenstern, in fact, very well interpreted the role of information in conflict situations and in the formation of coalitions; they did not take into account explicitly the cases of incomplete information or differential access to information under market conditions. They assumed that the rules of the game were known and that surprise moves would not take place. In reality, however, commodity markets such as that for coffee look more like chess games with several players negotiating in different dimensions; the content of the negotiations and the interaction among players changing through time. To add further complications, on many occasions the rules of the game are not clearly defined; that is a player may not have knowledge of the moves made by others and it may not even be clear which pieces the adversary has to hand.

It is interesting, further, to take note of the work of the Swedish economist, Stahl,[3] on the conditions under which the results of bargaining between two rational actors may be predictable. As would be expected, the assumption of rationality and complete information becomes crucial to determining the solution of the game on an a priori basis. Other authors such as Harsanyi[4] have worked on the concepts of power and incomplete information. Although there doesn't seem to be much literature on the role of information in commodity negotiations, it must be recognised that Keynes[5] was one of the first economists to consider that commodity inventories would tend to be suboptimal, given the uncertainty and the lack of information which prevailed in the market. This explains the background to his proposal to set up an international agency to manage stocks, and his support in the *Treatise on Money* of the Brazilian coffee valorisation scheme.

## Structure of the world coffee market

Coffee is the most important primary commodity in international trade in terms of sales after petroleum. Consumers spend more than 22 billion dollars on drinking coffee at the retail level. Coffee producing countries earn around 10 billion dollars – the difference being made up of tariffs, taxes, processing costs and profits. The crop is vital in terms of foreign exchange, employment, fiscal revenues and aggregate demand for a large group of Latin American, African and Asian countries. About 8.5 million hectares approximately are cultivated with coffee at the world level and 20 million people derive direct employment from its planting and initial processing.

Like most agricultural commodity exports, coffee shows low price and income elasticities of demand and unstable prices (see Table 5.1). Nearly 60% of its output is concentrated in five or six major producing countries, which have acted several times in a 'producers club' known as

**Table 5.1** *Characteristics of the world coffee market*

| | | |
|---|---|---|
| Price instability index[1] | 1949–1964 | 14.1 |
| | 1971–1976 | 23.7 |
| Hirschman producers' index[2] | | 0.09 |
| Hirschman consumers' index[2] | | 0.14 |
| Price elasticity of demand | | −0.4 |
| Income elasticity of demand | | 0.2 |
| Supply elasticities: | Short run | 0.2 |
| | Long run | 0.5 |
| Value of world coffee exports in 1980 approx. | | US $10,000 million |

[1] This index refers to average deviations from the trend.
[2] This index is defined as the square of the market shares of the agents participating in a given market; the higher the index, the higher the degree of concentration.
*Source:* O.I.C., World Bank, Colombian Federation of Coffee Growers.

the Bogota Group. This group was formally established in 1978 since it was clear that the main consuming countries were not interested in reactivating the economic clauses of the International Coffee Agreement (ICA). The main producing countries thought that the market could not be left to uncontrolled market forces; they therefore worked out a strategy at a 1978 meeting in Bogota aimed at establishing a stabilisation scheme within an acceptable range of prices. To achieve its objective, PANACAFE was legally established in Panama as an entity and endowed with the financial and technical capacity to intervene in the future markets of London and New York. This new company was funded with an initial capital of 140 million dollars.

This new institutional arrangement differed sharply from the multilateral export quota agreements which had been used since 1960 to regulate the market. The new organisation only involved producers: its main activities were concentrated in the future markets. It should be stressed that in the late 1970s export quotas were not in operation and various countries achieved record levels of external sales.

The degree of concentration in the international coffee market is higher among buyers than among producers (see Table 5.2). This is even more marked if one considers that green coffee processing is done by a very small group of multinational firms.[6] Processing also shows a marked tendency towards horizontal and vertical integration. A case in point shown by Pizano[7] is the Federal Republic of Germany wherein a group of six firms has virtually put out of the market nearly 500 smaller processing enterprises in the last twenty years.

The structure of the international coffee market may be characterised as a bilateral oligopoly. Such market structures work on the basis of a range of prices; the market solution depends on the bargaining power of the different participants. Radetzki[8] shows in a recent work that where

**Table 5.2** *Shares of major countries in the International Coffee Council*

| *Producers* | % | *Consumers* | % |
|---|---|---|---|
| Brazil | 21.2 | USA | 32.3 |
| Colombia | 15.9 | Federal Rep. Germany | 12.4 |
| Ivory Coast | 6.5 | France | 9.3 |
| Indonesia | 6.0 | Italy | 6.1 |
| El Salvador | 5.8 | Low Countries | 4.8 |
| Guatemala | 4.6 | Japan | 4.7 |
| Mexico | 4.5 | United Kingdom | 4.7 |
| Uganda | 4.2 | Canada | 3.2 |
| Kenya | 2.9 | Sweden | 3.0 |
| Costa Rica | 2.8 | Belgium-Luxemburg | 3.0 |
| | 74.4 | | 83.5 |

*Source:* The share refers to the number of votes each country has in the International Coffee Council. This figure is a bit different from the real market share due to the procedures used in determining voting capacity.

collective price elasticity of demand is lower, the exporting group will have a larger degree of negotiating power. In the case of coffee, producing countries have enjoyed a certain degree of bargaining power, particularly in circumstances where the disequilibrium between consumption and production has not been too large. But the oligopolistic power of roasters tends to work against this, especially when there are expectations of an increase in international stocks.

## Information access and the coffee negotiations

Differential information access becomes a crucial element of bargaining capacity in coffee negotiations. Information requirements and barriers will be considered in terms of international marketing, and international agreements. The major information variables to be analysed are of two kinds: volume and price data.

Volume information includes such items as output, consumption in producing countries, stock accumulation and levels in exporting countries, export and import volumes, consumption in major importing countries and stocks in consuming countries. In terms of reliability, the least reliable information typically comes from the producing countries and the most complete and up-to-date information comes from the major consuming economies.

Physical figures are crucial to both types of marketing arrangements. Within the International Coffee Agreements, the initial annual quota allotment is decided after a quantitative analysis of the coffee-year 'demand'. The latter is defined as the export shipments which would meet importers' consumption requirements, extrapolated from the previous year's figures, leaving the consumer only working stocks in hand. Leading producing countries usually favour low initial allotments, arguing their case in favour of relatively pessimistic consumption projections in terms of avoiding cases of over-stocking and the prevalence of large volumes of imports in transit. Symmetrically, consuming countries overstate import requirements and stress the need of guaranteed supplies. Such exercises are also included in producer arrangements negotiated between large world producers, seeking to reduce world supplies by spreading the stocking widely and the smaller producers, whose target is to export the entire year's output.

Volume data are also used to distribute the quotas amongst producing countries. Even though the International Coffee Agreements have semi-automatic clauses to establish the annual shares, countries which can argue on the basis of exceptional crops are able to obtain a readjustment of their shares. The ICA provides that 30% of the annual quota must be distributed on the basis of each country's share in stockholding, giving the output and stock figures a quite important role in international negotiations. For that reason such figures, especially the carrying stocks, become what one could call 'administered information', released only to top officials in producing countries, and subject to prior approval by the head coffee institutions in those countries. Such arrangements have led ICA authorities to promote coffee censuses in producing countries, and to seek the verification of stocks by specialised international institutions such as the General Superintendence Company.

In producer agreements such information becomes even more significant because past and prospective output figures set the base for distribution of the countries' shares of total sales; stock information being the control variable to verify if the countries are meeting their quota obligations. The traditional scheme followed in most producer arrangements has been to set up a 'retention' or stocking policy, whereby a given percentage of exports is stocked in order to regulate the market supply. Such procedures have promoted what may be termed 'the 200% annual output syndrome', the result of each producing country's output projections. This has tended to generate a world output estimate twice as large as the less unreliable US Department of Agriculture or International Coffee Organisation estimates.

Besides, producers' consortia have always had difficulty verifying the 'retention' efforts and penalising violations. In that sense, the value of International Agreements is that importing countries exercise quota controls by only accepting through customs duly stamped and certified exports. An example of producing countries' efforts to use the importers' information occurred when in 1973 a Producers' Group

(Geneva Group) decided to undership quota allowances and to base their joint effort on importers' statistical information and controls negotiated under the ICO. In such a way the exporting countries were able to introduce unilaterally their quota and price objectives within the ICO framework. The strategy did not last long since in the following coffee agreement two clauses were introduced to prevent such further actions. First, it was agreed that above a given price range quota allotments would not be made, and second, schemes to redistribute shortfalls were prohibited.

Such difficulties in establishing effective quota arrangements have led coffee exporters to design their strategies through price mechanisms. The Producers' Group activities in 1974 and 1975 as well as its actions in 1978, and 1979, were directed towards reducing the market supply by purchasing in the futures market and when necessary, taking delivery and stocking coffee. The strategy had a twofold purpose. By entering the future markets as buyers, producing countries hoped to increase the spot prices and, therefore, the selling price of their exports, independently of whether they obtained a pecuniary gain in each transaction as such. Secondly, such a strategy would allow the producers to end up collecting and storing any physicals (excess stock) overhanging the market. It was a classic developing country effort to try to enhance bargaining power by using industrialised country mechanisms and information. The main objective was not to maximise prices but to stabilise them within an acceptable range. The flow of coffee was not interrupted and coffee exports were at record levels.

Exporting countries almost completely lack information on the actions and policies of the multinational processing companies, which, because of their size and concentration, are able to influence market behaviour. The power of these corporations is reflected in the imposition of barriers to exports of processed coffee from producing countries. There is also widespread suspicion that they act collusively in the futures markets.

Another aspect of volume information which becomes crucial in international negotiations is the severity of Brazilian frosts. Statistical analysis of the frosts indicates that they are responsible for the medium run coffee price cycles and the severe fluctuations in short-run coffee price behaviour. The frosts bring about significant changes in income positions of the different operators in the market. Most traders, officials and representatives of other producing countries have strengthened their information sources on the possible incidence of frosts.

From the discussion of volume information, we now move on to a consideration of 'price information'. As in all commodities, there is a complete set of price quotations and indicators which must be considered when discussing price data. Although the range of prices varies from those at the consumer level to those paid producers in exporting countries, the most relevant quotations are those of green coffee at the processing level in international trade transactions. Green coffee prices also show a wide variety of quotations depending on factors such as the type of region and coffee trees grown (African robustas; Brazilian, unwashed arabicas; Colombian milds and other Central-American milds) and within such types even the particular country and region where it is grown (i.e. Mexican goodwashed; Guatemala's hard bean; Salvadorian high-grown). In addition, international coffee prices vary widely according to the place of transaction (New York, London, Le Havre, Hamburg and Breman) and according to the availability of the produce (i.e. spot New York, ex-dock, prompt shipment) and future quotations (the terminal London Coffee Terminal Market, the New York Coffee and Sugar Exchange and the future markets at Le Havre). In international negotiations a choice must be made as to the more relevant indicators to be used for purposes such as quota adjustments and price targets. Such choice is not, however, independent of the particular indicator chosen, since the functioning of the market is not good enough to set an adequate arbitrage among markets.

The International Coffee Organisation developed a system of daily indicator prices for each of the coffee types and a composite index which has served as a reference for quota adjustment. In the 1973 Agreement the composite index included the four main coffee types and in 1976, only the robustas and other milds were included as a result of the buying countries' insistence that the Brazilian and Colombian quotations were 'administered' by the officials in these two exporting nations. Such price indicators are obtained from market prices in New York as gathered and published by a private information broker, George Gordon Paton and Company and refer to ex-dock prompt shipment coffees.

Producing countries have been quite critical of the system used by the ICO authorities and have urged them repeatedly to use wider sources, in particular the quotations in European markets. 100 authorities reviewed the daily price information available in all markets concluding that overall the daily prices showed consistency both in levels and trends with other quotations. Some slight departures were, however, observed. Indicator prices tended to increase their differentials with regard to average import prices in the upward phase of the cycle; the 'other milds' indicator was also found to depart at certain points from the second and third positions in the New York and London markets, while a more consistent pattern was observed in the robusta's market. However, so far nothing has been done to improve price indicators.

Three major examples of the ways and means through which exporting countries have attempted to influence price behaviour by more skilled information handling may be identified. First, within the ICA, major producing and consuming countries sought to establish price differentiation through the selectivity arrangement, which linked the four types of coffees with corresponding price indicators in order to adjust quotas selectively according to the specific quality of coffee demanded in the market. Both Brazil and Colombia were thus able to readjust partially their quota allotments, given the power which they had to influence their own price indicators and given the weight which they had respectively in the exports of unwashed arabicas and Colombian milds. The reaction of the importing countries was to suspend the selectivity arrangement and the use of Brazilian and Colombian price indicators for quota adjustment purposes in the new ICA.

A second way in which the exporting governments' direct intervention influences coffee price formation is through the so-called surrender price scheme for minimum export prices. Such a strategy has been followed at times, mainly by Brazil, Colombia and some Central American and African countries which have an established exchange-control system. It requires exporters (even private dealers or transnationals) to deposit a minimum amount of foreign exchange at the central banks per bag exported, in order to obtain local currency exchange and forcing them to sell at minimum export prices. Such an action is usually coordinated by the producers' group or simply induced by Brazil acting as a price leader. It has become established practice that after a Brazilian frost all producing countries close their export sales; allow Brazil to adjust its surrender price; wait for indicator prices to move up and readjust their surrender prices to the new levels. In this manner, the commodity exporting nations prevent the multinationals from stocking up at cheap pre-frost prices, as was the previous practice.

A third way to boost prices directly has been attempted through the futures market. By coordinating actions to intervene in futures contracts, the Bogota Group sought not only to reduce market supplies and control the negative actions of speculators by purchasing futures and stocking coffee, but also to raise spot prices by altering expectations through the increased demand in futures. Such operations were financed partially by Euro-dollar loans. Producing countries have thus shown how to use information to their advantage in two formerly exclusive markets; futures and Euro-dollars. This is not to say that individual countries had not been active in the financial markets before. The Brazilian valorisation scheme of 1905, for instance, was financed partly by Rothschild, but coffee producers as a group had not made use of the Euro-dollar markets with the idea of financing futures market operations until the founding of the Bogota group. American response has been, as expected, to try to restrict producing countries' operations in the futures market, not only by altering the rules in the New York Exchange but also by pressurising the Bogota Group to stop consortium operations as a condition for a new International Coffee Agreement.

An obvious question arises at this stage: why did the Bogota Group suspend operations if it had been relatively successful in stabilising prices? The answer is relatively simple. 1979 began a period of over-production; it soon became clear that a more appropriate system to regulate the world's coffee market was the scheme of export quotas.

The financial requirements of future market operations in the presence of increasing stock levels were so large that producers opted for the International Coffee Agreement as a means of recreating order and stability in the world coffee economy.

## Conclusion

Using the example of coffee, this chapter has sought to indicate why the information access variable ought to be included in economic models of commodity markets and commodity negotiations. It is highly misleading to continue to apply the perfect competition model assuming total access to information, which has been the theoretical basis of most econometric exercises in the field of the simulation of the world's coffee market. The theory of dynamic games[9] although still in its infancy, might perhaps provide a better framework for understanding the complex process of bargaining which takes place in a bilateral oligopoly.

A very significant result of the coffee producers' experience is that the problem of unequal information access in commodity bargaining is not always apparent. In the case of the Bogota Group it has been shown that they developed the capacity to use to their advantage two sources of information previously reserved to the multinationals: the futures and the international capital markets. They also used to their advantage superior knowledge on two elements of the coffee market that became crucial to price expectations and price formation: the severity of Brazilian frosts, and (to a limited extent) stocks in the hands of producing nations. Finally, the Bogota Group obtained some short-term gains by exploiting to their advantage the structure of the ICA in items such as the selectivity arrangement and the use of import controls by industrialised countries.

Future international cooperation in the international coffee market ought to be based on consumer collaboration, according to certain international coffee leaders. They argue that their help is needed in order to ensure control of the system of export quotas, and that all participants benefit from stability in the world's coffee economy. The ICA offers them regular supplies at stable prices. Simultaneously, consumers understand that stable and remunerative coffee prices maintain the level of the world's coffee supply. This enables the producing countries to continue import levels of manufactures and capital goods from OECD countries at a significant level.

In 1982 delegates from 42 producing countries and 20 consuming countries defined the extension of the ICA for the rest of this decade. Most experts and observers think the IEC will be extended with some adjustments. If this is the case, it does not appear likely that the Bogota Group will be active in the immediate future. Nevertheless, in this particular case, producers have shown that they are far from impotent and powerless in their dealings with multinational corporations. The coffee market will not, however, be very easy for producing countries during the 80s. It is likely that coffee production will continue to outpace demand through the decade; prospects for increased consumption in key markets such as the US are not very favourable. It is possible, however, that the new ICA could help ease the problems associated with overproduction and chronic instability which have been so difficult for Third World coffee producers to date.

## Notes

1 K. Arrow and F. Hahn, *Análisis General Competitivo*, Fondo de Cultura Económica, Mexico, 1977.

2 H. Von Neumann, and O. Morgenstern, *Theory of Games and Economic Behaviour*, Princeton, 1944.

3 Stahl, *Bargaining Theory*, Stockholm, 1973.

4 J. L. Harsanyi, 'Games with Incomplete Information played by Bayesian Players', *Management Science*, 1967.

5 J. M. Keynes, *A Treatise on Money*, Collected Works, Macmillan, London, 1971.

6 A. Gordon, 'The Location of Coffee Processing', OECD, Development Centre, 1979.

7 D. Pizano, *Procesamiento y Consumo de Café en la República Federal de Alemania*, Bogotá, Fondo Cultural Cafetero, 1981.

8 Radetzki, 'Monopolistic Commodity Pricing' in G. K. Helleiner, (ed.), *The Less Developed Countries in the International Economy*, Cambridge, 1976. Radetzki assumes that this elasticity (Edc) can be defined as follows:
Edc = 1/M. Edw – (1/M). (1–M). ESr.
Where M = share of the group in total world supply
Edw = world demand elasticity for the product
Esr = price elasticity of supply for producers outside the cartel.
It must be pointed out that this is a necessary but not a sufficient condition for a cartel's success. One must also take into account income elasticity of demand and the counter-strategies designed by the consumers.

9 Kalai, Maschler and Owen, 'Asymptotic Stability and other Properties of Trajectorics and Transfer Sequences Leading to the Bargaining Sets', *International Journal of Game Theory*, 1976.

# 6

# Marketing, Technical and Legal Information for Copper Producing Countries: The Case of CIPEC

Carlos Fortin

## Introduction: CIPEC and the world copper industry

The Intergovernmental Council of Copper Exporting Countries (CIPEC) was set up in 1967 by the governments of the four largest copper producing countries of the Third World – Chile, Peru, Zaire and Zambia. Since then it has operated continuously and its membership has been extended to include Australia, Indonesia, Mauritania, Papua New Guinea and Yugoslavia.

At the time of the establishment of CIPEC, production and marketing of copper in all four founding countries were effectively controlled by foreign companies, despite the formal nationalisation of the Union Minère operation in Zaire in January 1967 and the acquisition by the Chilean government of a 51 per cent interest in the equity of one of the two US companies in Chile (a Kennecott subsidiary) in April that year.[1] By and large, the governments were not prepared to interfere with the running of the industry by the foreign firms, and therefore the new organisation was essentially conceived as a forum for the exchange of views on broad policy issues – a role whose limitations were heartily endorsed by the multinationals themselves.

The longer term objective, however, was to coordinate measures to increase the countries' earnings from copper exports and to harmonise policies relating to the production and marketing of copper.[2] In the early 1970s it appeared as if the objective was close at hand. This was a period of rising nationalism and self-affirmation in CIPEC countries. In Peru the military government that took over in 1968 nationalised the International Petroleum Company (IPC), a subsidiary of Standard Oil of New Jersey, and went on to reserve for the state the international marketing of Peruvian copper in 1971 and to nationalise one of the two US-owned copper operations, Cerro de Pasco, in 1973. In Chile, the gradual takeover of the large copper mining sector initiated in 1967 was dramatically stepped up when Salvador Allende fully nationalised the operations of Anaconda and Kennecott in 1971. In Zambia, joint venture arrangements were introduced in 1969 whereby the state acquired a majority of the equity in the foreign companies operating the copper mines, but the foreign partner retained control over management and sales; this soon proved less advantageous for the country than had been expected, and the government's dissatisfaction culminated in 1974 in the decision to terminate the arrangements involving foreign responsibility for management and marketing with the functions reverting to the Zambian companies. The one exception was Zaire, where for the first three years of the decade the copper industry and copper policy were clearly dominated by the Belgian Société Générale de Minerais. But even there some increase in government involvement occurred along with the gradual emergence of self assertive attitudes: in June 1974 the so-called Munich Agreement between the Zairean Government and Société Générale de Minerais led to the creation of a Zairean marketing company and the partial establishment of Zairean control over sales.

The promise, though, has hardly been fulfilled in the intervening decade; CIPEC countries have not increased substantially their control over the price of their copper, nor have linkages among their copper industries of their economies developed significantly. There are several reasons for this. The paramount factor is that throughout CIPEC's history there has always been at least one government which values good relations with international capital more highly than collective self-reliance among the member countries. Until the mid-1970s the country was Zaire. This became apparent when a dramatic drop in the level of prices which took place in the second half of 1970 led to discussions about a possible intervention in the market to halt the decline. The Allende government in Chile was a main promoter of this initiative. The discussion was carried out among the operating companies of the CIPEC countries, and by request of Société Générale de Belgium, CIPEC as an organisation was excluded. The proposal was for the introduction of a floor price coupled with the establishment of a buffer stock to be financed by the producers, that could absorb the excess supply for a period and then release it when prices went back to normal levels. The proposal foundered on the question of how to apportion the cost of the operation. In the process it transpired that the Société Générale was less than enthusiastic about embarking on the operation at all, and as they were for all intents and purposes speaking on behalf of the Zairean government the matter was effectively shelved in mid-1971.[3] A similar attempt in 1974 – this time a production cut-back – was somewhat more successful, but it fell short of a sustained and effective intervention.

After 1974, the member country which had been hostile to proposals for intervention in the market was Chile, following the 1973 coup that overthrew the Allende government and installed a military regime bent on attracting foreign capital. The possibility of a second production cutback came up in late 1977, and this time it was Chile which decided to exclude itself from the initiative. This did not prevent the cutback from being further considered and later – in March 1978 – adopted by Peru, Zaire and Zambia. The May invasion of the Shaba province by Katangese forces and the subsequent disruption of production in Zaire, which raised the international price of copper, made implementation of the cutbacks unnecessary. It is not clear, though, whether the three countries would have gone ahead without the participation of Chile, the largest producer.

Another factor militating against the involvement of CIPEC as an organisation of joint action by members' countries is the enlargement of the membership of the organisation. Expansion beyond the original founding members was justified in terms of the increased monopoly power that a larger membership would entail. On the other hand, the problems, as suggested, have had more to do with the diversity of broad policy orientations among the members than with a specific calculus of costs and benefits, and the enlargement could only enhance heterogeneity. The only new full member, Indonesia, is not a large enough producer to make much of a difference. The two large producers which joined in 1977, Australia and Papua New Guinea, are associate members. Australia is, of course, a developed country with a mining industry which is wholly operated by private companies; Australia has made it clear that it is not prepared to participate in any decision that may require the dictation of terms to the private companies. It has more or less consistently opposed the Third World proposals in UNCTAD. Papua New Guinea appears less than interested in the possibility of market intervention as does Yugoslavia, a semi-developed socialist country. Finally, Mauritania is a very small producer, for which the copper industry is a relatively minor part of the economy.

In view of these factors, it is not surprising that CIPEC has not become a major actor in the world copper industry. Conversely, from the beginning it was recognised that CIPEC could help improve the knowledge of the governments about the industry and the market through the generation and exchange of relevant information. Thus, the permanent body of the organisation was a Copper Information Bureau, later renamed Information Secretariat. Its brief was defined as providing information and submitting reports on demand, supply and marketing of copper; on governments' agreements and regulations affecting the industry; on technical advances in mining, processing and use of copper; and on labour, transport and other economic factors in the industry. The Bureau consisted of an Executive Committee composed of representatives of the member countries, of an Executive Director, and of a staff of one expert seconded by each of the member countries plus one statistician, one economist, administrative personnel and translators. It was conceived as an independent technical body; the CIPEC Agreement states that the members' countries 'shall respect the exclusively international character of the responsibilities of the Executive Director and staff and shall not seek to influence them in the discharge of their duties'.[4]

The need for a source of information and intelligence which was independent of the copper multinationals was common to all four governments, although their capabilities in terms of data gathering, processing and utilisation was variable: Chile was probably the most advanced in this respect, with Zambia and Zaire, on the other hand, being almost completely dependent on the foreign companies for information concerning the industry. This need was intensified in the early 1970s with the increased involvement of the governments in the countries' copper industries: as the governments' intervention in their copper industries and markets expanded, so did their needs for accurate and timely information for decision-making. What kind of information was required and for what specific purposes? Before attempting a preliminary answer to this question a brief description of the world copper industry and market is in order.

Raw copper production involves basically four technical operations, each of which results in a particular kind of product for which there is a market.[5] The first one is extraction, resulting in copper ores with a copper content that varies depending on the grade of the mineral: the range of variation of major operations goes from a maximum of about 6 per cent of copper content in Zaire to about 0.35 per cent in some US mines. In any case, the high transportation cost of the ores makes them at less than suitable material for international trade, and consequently its separate market is very small.

The second operation is concentration, where ores, after being crushed, are floated in a chemical solution in order to separate the copper-bearing particles from the waste material; the result is copper concentrate, with a copper content of 20 per cent to 50 per cent for which there is a substantial international market

The third operation is smelting, where concentrates are treated first in a reverberatory furnace and then in a converter to produce what is known as blister copper, with a typical copper content of up to 99.5 per cent. Finally, the blister copper is refined by means of electrolysis to produce a material of 99.92 per cent to 99.96 per cent purity, which may be in the form of thin sheets of metal known as cathodes or shaped into wirebars or other forms. Electrolytic copper is the standard form of copper traded.

The consumers of copper are fabricators who process raw copper into standard shapes, sizes, temper and finishes required by manufacturers of finished products – such as electrical cable, radiators – or by construction industries. The main fabricator products are wire, tube, pipes, sheet and plate and rod. As copper is, for all practical purposes, an indestructible material it is possible to reprocess copper and copper alloy products to recover their metal content. Copper scrap is thus an important source of metal alongside primary copper.

The world's two largest producers of copper are the United States, with about 22 per cent of world capacity and the Soviet Union, with 14 per cent. On the whole they both consume as much copper as they produce, the US being a small net importer of refined copper, the USSR a small net exporter. The largest exporters of copper are Chile, with 12 per cent of total world capacity, Zambia with 9 per cent, Zaire with 6 per cent and Peru with 4 per cent (together with Canada which has 9 per cent of world mine capacity). All four main CIPEC countries sell the bulk of their copper through direct contracts with foreign fabricating firms, mostly in Europe, Japan and the US. The contracts are usually for one year and establish a total tonnage to be delivered in monthly instalments. The negotiation of the contracts is done once a year, normally in October and November. The price paid by the buyers is determined by reference to the London Metal Exchange, a futures market established in 1881 and organised as a private company composed of some 100 member firms, although only some 28 of them are entitled to buy and sell on the floor of the Exchange. The amount of physical copper traded in the LME is small and none of the main CIPEC countries as a general rule sell their copper in it. Nonetheless its official daily price serves as the basis for the price charged by the CIPEC producers; essentially they charge the average price of the month of delivery. The price for unrefined copper is determined by deducting from the electrolytic quotation the smelting or refining charges. The level of the latter is a matter for negotiation and depends on the particular characteristics of the material sold.

## The information needs of Third World producers

In the early 1970s the governments of the CIPEC countries were faced with the need to adopt policies and make decisions on several crucial areas regarding their copper industries:

1 the evaluation and redefinition of the regime of operation of the existing foreign companies, including taxation, foreign exchange regulations, depreciation allowances, input procurement, manpower training and deployment, service fees, etc.; and covering both their involvement in productive activities and in marketing;
2 the regime for new foreign investment in the copper industry, including the negotiation of specific concession agreements, where the issues are basically similar to those suggested above;

3 the expansion of State owned activities, including prospection and exploration for new mines, expansion of existing extractive operations and vertical integration into smelting, refining and fabrication;
4 the adoption and implementation of policies concerning the running of the nationalised productive operations with a view to minimising the difficulties of the transition from private foreign control to state management; this included problems of depletion and reinvestment, of sources and terms and conditions for the procurement of inputs, intermediate products, capital equipment and spare parts, and of labour relations;
5 the adoption and implementation of sales policies, including decisions as to what kinds of products to sell, to whom and on what terms and conditions; and
6 conceivably, the joint intervention of CIPEC producers in the market in order to support prices or to stabilise them around agreed levels.

It is immediately apparent that in the preceding policy-making areas, specific types of information and knowledge are required, with respect to various aspects of the industry. The kinds of information required are: *legal*, including both questions of internal and international law; *technical*: geological, extractive, metallurgical; *economic*, concerning the economics of production, of taxation, of foreign exchange regimes; and *commercial*: markets and marketing terms and conditions, as well as the economics of trade. As for those aspects of the industry on which information is required, there is first, the structure of the industry and the market and trends in their development. Technical and economic information about levels of existing and planned world-wide productive capacity by types of productive operation, and legal information about patterns of ownership and control are indispensable data for negotiation with existing or prospective investors: (a) to decide on production and investment policies affecting the output mix; and (b) to decide on a sales policy, to bargain with specific buyers. An overall grasp of the structure of the industry world-wide is a precondition: it must be susceptible to frequent up-dating in order to identify trends. Equally, knowledge is required about existing and emerging technologies for exploration, extraction, processing and fabrication, as well as the structure of the world market for equipment and inputs. A sales policy or, more importantly, a policy of price intervention requires an understanding of the basic dynamic of supply and demand including: (a) structural determinants of demand and their tendencies; (b) the structure of world transport and of the operation of specific market mechanisms, such as futures markets. Indeed, reliable and unbiased information about the structure of the operation of the country's own mining sector as well as its markets is necessary for dealing with the first major issues specified in the beginning of this section.

A second type of required information concerns flows. Its central component is commercial information about short term movements of demand and supply for each of the products described above, and the scrap market; about variations in stocks; and about the day to day operation of the futures markets; but the essence of the requirement has to do with the timeliness of the information. It may be argued that this kind of information is less relevant to the needs of copper exporters given that, as suggested above, the bulk of their sales are through annual contracts at a price determined by the London terminal market. The CIPEC countries cannot modify their pattern of sales at will to take advantage of short-term market fluctuations. On the other hand, there is always a certain element of flexibility in that a proportion of the expected annual output, which in fact has been growing, is not sold through annual contracts. Furthermore, information about flows would become crucially relevant should there be an attempt at changing the pricing system or intervening in the market to support or stabilise prices.

Finally, there is information concerning policies, be it of the Third World copper producing country governments, of the governments of consuming countries or of the relevant multinational corporations. Legal knowledge about the foreign investment regime in Third World producing countries and the changes it undergoes, economic information about the accumulated experience of the different governments with various sets of regulations, and bargaining with foreign companies are valuable as contexts for comparisons and as lessons. This also extends to commercial information concerning terms and conditions in current sales contracts and the distribution of markets. The policies of the advanced consuming countries on tariff and non-tariff barriers to processed copper, on stockpiling, on subsidies and internal taxation to local industry, and on future development, including technological development, must be taken into account when deciding on policies on the expansion of production or vertical integration. Medium and long-term corporate expansion plans are also crucial informational inputs for such decisions.

Some of the information mentioned above is relatively easily accessible in raw form, and does not require highly specialised knowledge to be processed and analysed; what is required principally is knowledge of the sources and a mechanism to tap them on a continuous basis. Thus, for instance, information about productive capacities and about ownership and control can be obtained from industry and trade journals, company annual reports and specialised publications. In other cases, information is available already in initially processed form, including production figures and trade flows, which are recorded and publicised monthly by the World Bureau of Metal Statistics, a UK-based organisation financed by producers and consumers of copper, as well as by several US organisations. Their analysis requires no more than basic statistical and econometric skills. In other cases, though, information is either not available easily, or requires highly specialised knowledge for interpretation and analysis. Details of commercial transactions are generally confidential, as are most operations in the London Metal Exchange. Insider access to the firms, governments or institutions involved is necessary to acquire knowledge about them. On the other hand, technological data, sophisticated econometric models or complex legal arrangements call for highly specialised training and skills in order to be of use for decision-making, even assuming that they might be publicly available.

All the types of information summarised above are as relevant today to the needs of CIPEC countries as they were in the early 1970s. Governments of CIPEC countries have some capabilities in respect of all of them, but this usually falls short of what is required for optimal decision-making. Even the Chilean state copper organisation, generally acknowledged to be one of the most professionally advanced among CIPEC countries, does not seem able to handle the informational aspects of some negotiations. In a recent negotiation with Exxon concerning the concession to exploit a major copper deposit, for example, the Chilean government's estimate of the copper content of the deposit was 400,000 tons. After the concession had been agreed in highly favourable terms for the company, Exxon announced that subsequent geological work had revealed the reserves to be really 6 m tons of copper.[6] It is possible that the company was aware of this from the start of negotiations. Indeed it has been suggested that the Chilean government also suspected the original estimate to be too low, but refrained from pursuing the matter, in order not to prejudice the negotiations. This would have been very much in line with the present government's general policy of attracting foreign investment at any cost. It seems clear, however, that adequate technical advice might have prevented an agreement which was clearly more advantageous to the foreign company than the situation warranted. Has CIPEC played a role in supplementing the informational capabilities of the member countries to any significant extent? Can it play such a role in the future? The two questions will be discussed in turn.

## CIPEC's performance in the field of information[7]

The first task that CIPEC undertook in the information field was the building of a statistical data bank and the production of quarterly reviews of the copper market. The data collection did not go beyond what was already available in the organisations mentioned above, but the regular reviews, although, again, not going much further than the offerings of existing commercial copper information services, made easily accessible to member governments the current state of thinking on the evolution of the copper market. As suggested above, this was of particular interest to the member countries whose own data analysing capabilities were limited. CIPEC was in effect providing technical assistance to the least developed members in the form of market data and analysis. A comparative study of the mining legislation of the member countries, later extended to the other major producers, was also undertaken. By far the most substantial studies, however, were commissioned by the organisation to outside experts. A study of the operation of the London Metal Exchange was carried out by the Commodities Research Unit of London in 1969 and a study of future copper demand, supply and consumption was done by the Battelle Institute of Geneva in 1970–71. They, no doubt, contributed significantly to the understanding of the world copper market among members, although, again, it was those governments which already had a more developed data gathering and processing capability which could take full advantage of the data contained in the studies.

In contrast to the modest but not inconsiderable progress made in the areas of legal and economic statistical information, CIPEC found it impossible at first to tackle information issues relating to commercial policy. Early suggestions for the exchange of information on the terms of sales contracts of the member countries were resisted by Zaire, which argued that the contracts were a matter of commercial confidentiality and that the countries, even though having some common interests in the market, were essentially commercial competitors. Thus, the Chilean proposal to create a sales committee within CIPEC, which had been originally agreed in the 1967 founding conference, was opposed by Zaire at the meeting of the Governing Board of May 1968 and was effectively shelved for the following five years. All through this period, therefore, CIPEC was prevented from having access to the clauses of sales contracts of the member countries, and there was no exchange of such clauses among the countries themselves either.

This essentially reflected the distrust that the foreign companies in control of production and sales in the member countries had of the new international organisation. Commercial confidentiality did not however, prevent the companies themselves from having regular annual meetings (to which private companies operating in countries other than CIPEC members were also invited) or where the basic clauses of the sales contracts were discussed and agreed.

The situation changed with the creation in 1973 of the Permanent Committee on Marketing Policy. This move, adopted at a time of self-assertiveness in all member countries, initiated a process of increased exchange of information which has continued despite subsequent reversals towards a less cooperative attitude. An important expression of this progress was the 1978 decision by CIPEC members jointly to introduce some modest (though not negligible) changes in the pricing terms of their contracts. By contrast, the existence of CIPEC has had no impact on the actual selling process and mechanisms in the member countries. There have been no moves towards coordinating their sales organisations or unifying actual bargaining with the consumers, which would have obvious commercial advantages. In this area the perception of other members as competitors still dominates the thinking of all CIPEC countries.

Another area in which CIPEC has endeavoured to improve information for policy-making concerns the possibility of integration of the copper industries of the member countries. Statements emphasising the importance of cooperation in manpower training, geological and technological research and production, and the exchange of mining equipment and inputs date as far back as 1969. The Governing Board meeting of May 1970 requested countries to provide information on the input requirements of their mining operations in order to begin studying the possibility of joint production of some of them. Discussions led to several studies: a survey of mining equipment and other inputs with a view to possible standardisation; a study on access to markets for new products, such as continuously cast rods; and consideration of the creation in 1973 of the Permanent Technical Committee charged with continuing discussion of possible technical cooperation.

The actual impact of these studies and discussions has, however, been negligible. CIPEC's Secretariat simply does not have the technical capacity to produce or retain information which can provide the basis for action in this area. Furthermore, the member governments seem to have decided in favour of further integration of their copper industries with the advanced countries' copper economies. Thus, Chile, Zaire and Zambia have entered joint ventures with European firms to set up continuous cast plants for the production of rod located in Europe, to which a regular flow of refined copper from the countries is assured. This trend is a function both of the specific preference for advanced technologies which still prevails in the thinking of copper decision-makers in the producing countries, and the more general trend in their governments towards a firmer insertion of their economies in the world capitalist system. This is, of course, an area in which significant potential exists for a gradual process of delinking of the producing country copper industries from advanced country sources and relinking them within the CIPEC group of countries. Imported inputs, notably chemicals, machinery and spare parts, make up a substantial proportion of production costs. Efforts to coordinate the procurement of inputs, in order to improve the collective bargaining position, and ideally in some cases to replace imported items with jointly produced local ones, are an obvious way of reducing costs and/or external vulnerability. Integration of some production facilities, such as energy sources or even smelting or refining plants, is conceivable for Latin American producers as a group and for African producers, as is the planning of joint ventures in downstream integration such as rod production. The central problem, though, is less one of information than of political will.

A third area in which CIPEC has done work, but has been unable to make a major impact, is that of bargaining with prospective international investors. Although, as suggested, the organisation has carried out comparative studies of the regime offered to foreign investors, member countries have not felt it necessary to consult CIPEC sources for advice in particular bargaining contexts. They have essentially relied on private advisers or on organisations such as the Commonwealth Secretariat. Occasionally they have requested assistance from other individual member countries. This reluctance may reflect the misgivings of member governments to allow confidential and sensitive negotiations to be open to representatives of an intergovernmental Third World organisation. The expansion of CIPEC referred to above has in all likelihood increased these fears and made it further unlikely that CIPEC will be regarded as a major source of advice on negotiations with multinationals by its constituent governments.

CIPEC has finally carried out studies on the possibility of introducing a producer's price which, although still linked to the LME price, would allow the member countries a higher level of control over the pricing of their copper, while at the same time having a stabilising effect. This is linked with the more general question of the possibility of CIPEC producers deploying a cartel strategy or, alternatively, collaborating with the consuming countries, in order to achieve some degree of price stability. Until 1975, discussions in CIPEC centered around the possibility of unilateral intervention to support the price in periods of slack demand; in 1974 an attempt was actually made by means of cutting back production by 10 per cent. The move was generally credited with having had a modest impact on the market, i.e. it probably prevented the price from falling further. On the other hand, its original target was not sufficiently high, it was not fully implemented by all countries, and, most importantly, it reflected a political situation in CIPEC countries – of nationalism and self-assertiveness – that was not to last. By 1975, both Chile and Zaire moved away from unilateral action: the Lima Conference of Ministers of CIPEC in December 1975 formally abandoned the unilateral approach in favour of trying to reach agreement with consumers to stabilise the price through the operation of a jointly administered and financed buffer stock. As the initiative in this direction was seized by UNCTAD at its Nairobi conference in 1976, the latter effectively replaced CIPEC as the arena in which reform of the world copper market mechanism was being discussed. While CIPEC continued to attend the various preparatory meetings called for by Resolution 93 (IV) of UNCTAD, the position of the member countries was presented by their own delegations, and there was little previous coordination of them through CIPEC. CIPEC was not called upon to provide technical or informational support to the member countries.

Once again, it would appear that the relegation of CIPEC to a secondary position is a matter of political decision. On the other hand, UNCTAD began to produce a steady flow of in-depth studies of the copper market, including some commissioned to outside consultants, thus making CIPEC involvement still less necessary (CIPEC itself has contributed 3 studies: a brief description of marketing and pricing mechanisms; an estimate of capital requirements and probable costs of planned new copper operations; and a summary of copper-trade methods and figures for 1975).[8]

## Improving the information capacity of CIPEC

How much further can CIPEC go in performing a useful function as a central generator, repository, processor and disseminator of relevant information for member countries? If our preceding analysis is correct, it would appear that the prospects are not promising. A combination of improved information at the world level, including CIPEC's own success in generating a flow of basic data, and of strengthening of the informational capabilities of the member countries, means that, at least for the founding countries, CIPEC has become increasingly less relevant as a source of statistical data and analysis. This, of course, does not necessarily apply to new entrants in the world market, such as Indonesia and Papua New Guinea (CIPEC members) or the Philippines (not in CIPEC). But these countries have so far shown only limited interest in expanding CIPEC informational capabilities at this basic level.

It would be conceivable for CIPEC to expand qualitatively into more complex informational areas, such as market management, technological development, or project evaluation, which would, of course, require significant expansion and strengthening of the organisation's structure and staff. But such a development is only possible if there is a shared political option to reinforce the links among Third World producers at the expense of links with international capital. As suggested above, the opposite trend seems to prevail at the moment in a majority of Third World copper producers. To take only CIPEC's founding members, it seems clear that the trend is the enthronement of politico-economic projects, which are essentially based on a growing integration in the world capitalist system. The most extreme case is that of Chile, where the military regime's economic model is one of opening the economy fully to the world capitalist market. Zaire also, less by design than as a result of a deep and growing economic crisis, has become increasingly dependent on, and integrated with, Western capital and finance. Peru, again, partly as a result of serious internal economic imbalances, is gradually abandoning the more self-reliant approach of

**Table 6.1** *CIPEC member countries: some selected indicators*

| | *GNP US$m* (1) | *Per capita GNP US$* (2) | *% GNP from agriculture* (3) | *% GNP from industry* (4) | *Copper revenue as % of total export revenue* (5) | *Refined copper production as % of total copper production* (capacities) (6) | *Cost of production 1977 US ¢ per pound* (7) |
|---|---|---|---|---|---|---|---|
| Chile | 23,980 | 2,160 | 7.9 | 21.4 | 45.7 | 60 | 40 |
| Peru | 16,470 | 930 | 12.6 | 24.5 | 19.2 | 60 | 49 |
| Zaire | 6,340 | 220 | 26.1 | 7.2 | 35.3 | 50 | 55 |
| Zambia | 3,220 | 560 | 12.3 | 11.7 | 95.6 | 95 | 59 |
| Indonesia | 61,770 | 420 | 32.3 | 12.9 | .005 | 0 | n.a |
| Australia | 142,240 | 9,820 | 4.9 | 20.2 | .02 | 98 | n.a |
| Mauritania | 530 | 320 | 2.6 | 2.3 | – | 0 | n.a |
| Papua New Guinea | 2,360 | 780 | 29.7 | 7.5 | 51.1 | 0 | 54 |
| Yugoslavia | 58,570 | 2,620 | 13.7 | 38.4 | .02 | 100 | n.a |

*Sources:* (1), (2) *World Bank Atlas*, 1981.
(3), (4) *Economist Intelligence Unit Quarterly Reviews*, Annual Supplements: figures for Chile, 1980; Peru, 1979; Zaire, 1977; Zambia, 1980; Australia, 1977–78; Indonesia, 1979; Mauritania, 1979; Papua New Guinea, 1975; Yugoslavia, 1979. Figures for industry in Mauritania and Yugoslavia include mining.
(5) Ibid. for Chile (1980), Mauritania (1980), Peru (1979), Zaire (1979), Papua New Guinea (1979); for Zambia, *Zambia Mining Yearbook* 1980; for Indonesia, Australia, Yugoslavia, UN *Yearbook of International Trade Statistics*, 1979.
(6) C. Fortin, *The Political Economy of the World Copper Industry*, 1980, pp. 30.
(7) World Bank, Chile, An Economy in Transition, 1980, p. 211.

the Velasco government and adopting the kind of economic recipe for financial stability proposed by the International Monetary Fund, which entails liberalisation of foreign trade and a reduction in the role of the State in the management of the economy. Zambia is thus the only founder member still broadly committed to a strategy of collective self-reliance. As suggested above, the enlargement of the organisation has only compounded the political difficulties of moving towards true Third World cooperation among CIPEC members. In addition, differences in the economic and industrial structure of the countries and particularly in the costs of production of copper, already present in the four founding members, have been intensified. The magnitude of the differences among the countries can be seen in Table 6.1, containing figures on selected indicators.

In conclusion, it seems clear that while CIPEC as an organisation can – and probably will – contribute to increasing interaction, mutual knowledge and exchange of views among member countries, the real chances for a stepping up of collective self-reliance among them are more problematic. It has been argued that these factors are a function of the models of society which the ruling groups in these countries adopt and implement. The present models run precisely counter to the notion of the Third World emerging from dependency by increasing cooperation and solidarity and, while they prevail, it is unlikely that CIPEC will be called upon to enhance and strengthen its informational and technical capabilities.

## Notes

1 The companies were: Anaconda and Kennecott (US) in Chile; Southern Peru Copper, a subsidiary of American Smelting and Refining Co., and Cerro Corporation (US), in Peru; Société Générale de Minerais (Belgium) in Zaire; and Anglo American Corporation (South Africa, UK) and Roan Selection Trust (US, UK) in Zambia.

2 *An Agreement to Establish the Intergovernmental Council of Copper Exporting Countries*, 1967, Art. 2.

3 C. Fortin, 'Third World Institutional Politics and Collective Self-Reliance: The Intergovernmental Council of Copper Exporting Countries', in E. Oteiza, C. Kenrick and J. A. Zammit (eds), *Collective Self-Reliance: An Alternative Development Strategy*, forthcoming.

4 *An Agreement* . . . , art. 17.

5 This highly simplified description refers solely to copper contained in sulfide ores, which are the more common. For a detailed treatment see A. D. McMahon, *Copper. A Materials Survey*, 1965.

6 C. Vignolo, 'Inversión extranjera en Chile 1974–1979. Una evaluación crítica', *Mensaje* (Santiago, Chile), No. 286, enero-febrero 1980, p. 40.

7 This section is based on C. Fortin, op. cit.

8 UNCTAD DOCSTD/B/IPC/COPPER/AC/L 9, L 10 and L 11, 1977.

# 7

# Landsats, Minerals and Development: A Qualitative Notion of the Down-Side Risk*

William Lazarus

## Introduction

Remote sensing satellites make detailed images of the earth's surface in digital form, which can be transformed into maps providing information on, among other things, the environment and weather, crop yields and diseases, and mineral deposits. The United States is the leader in this field with its Earth Resources Satellite Programme, started in 1972. In principle, raw data from its Landsat is available to any nation, but in fact the differential capacity to use and programme the data for specific purposes is highly unequal. Landsat data, processed with a particular interest in geological data, can be of valuable assistance in detection and quality assessment of mineral deposits, particularly in developing regions. Geologists of most of the major mining corporations are now able to check their assumptions concerning deposits against data from the satellite, whereas the capacity to use or interpret this data among mineral producing countries of the Third World is decidedly more limited. The use of remote sensing devices is not, therefore, regarded by all nations as an unmitigated global asset.

In political terms, the international debate on the regulation of satellite remote sensing (a debate focused primarily, but not exclusively in the UN Working Group on Remote Sensing of the Earth by Satellite) has generated three main approaches.[1] The first holds that a special treaty ought to be required to govern remote sensing from space, and that the treaty must guarantee the sensed nation the right to consent to be sensed, and to control the dissemination of data acquired (it has been

* This chapter was available in an earlier version as an Occasional Paper of the Research Programme on Communications Policy, Center for International Studies, Mass. Institute of Technology, Cambridge, Mass., 1981. It has been revised by the editor on the basis of updated material submitted by the author.

supported by Brazil and Argentina in international debates).[2] The second is a French-Soviet Union position which proposes a set of guidelines, rather than a formal treaty, which, while not providing the sensed nation the right of prior consent, allows it to control the dissemination to third parties of data acquired from space. The third is the US position consistent since the start of the debate which maintains that the 1967 UN Outer Space Treaty is a more than adequate legal framework within which to place remote sensing from space, requiring no further treaty amendment. This position is based on the assumption that open dissemination of data under the existing system is in the best interest of everyone concerned.[3]

Many scholarly works on remote sensing consider its international legal implications,[4] while the literature on remote sensing for development may be very roughly categorised under three broad rubrics: firstly, technical works on issues or problems of a specific scientific or engineering interest; second, reports of field projects and efforts in technology transfer; and third, cost-benefit forecasts or works on the potential or prospects for the technology in developing nations. Very few works, indeed, consider political and economic implications. The cost-benefit literature indicates that the potential benefits of remote sensing in support of development are considerable, and technical and field studies indicate that progress is being made at all levels in applying remote sensing from space to development problems. There remain, however, long-standing disagreements on prior consent and access. Take, for example, the case of Brazil (one of the most advanced of the developing countries) with considerable experience and investments in satellite remote sensing,[5] which has opposed open access and dissemination of data from their country.

The international debate can hardly be dismissed as *merely* legalistic, though there is surely no lack of hairsplitting in the various proposals and counter-proposals in the UN Working Group, which have been characterised by deadlock for a decade. The debate on remote sensing reflects the politicisation of what some regard as purely technical issues; but to consider it only as political posturing devoid of significant economic substance would be a grave error. 'Behind the legal disputations over sovereignty lies a fear among some developing countries that remote sensing technology might confer advantage less to themselves than to other parties . . . especially the oil- and mineral-extraction companies and agricultural commodity traders.'[6] That a number of developing countries *believe* that Landsat data about them will redound to their economic disadvantage, is unquestionable. What is questionable is whether there exists any realistic economic basis for the fear.

The 'prospects and potentials' literature tends to mention the fears and to suggest that they are not economically warranted.[7] Some of the legal literature tends to take the fears more seriously. Citing a number of important precedents from the US and Western Europe, one author[8] suggests that there are striking analogies between 'unjust enrichment' from unauthorised aerial surveys on the one hand and satellite surveys on the other. The technical and applications literature tend to be silent on the subject.

The assumption of most of the published work on remote sensing, apart from a few UN documents and legal articles, is that satellite remote sensing is a great potential boon for development. My research has led me to conclude that the blessing may be a mixed one; that the fears reflected in certain developing countries push for a more 'restrictive approach' to remote sensing are not chimerical but real in at least one important sector – the use of satellite imagery for minerals and oil exploration/extraction.

At some future time it may be possible to develop a quantitative model of the 'down-side risk' associated with asymmetrical capabilities of multi-national minerals corporations and developing countries to use Landsat data in minerals negotiations. The objective of this chapter is much more modest: to gain a qualitative understanding of the factors which lead to an asymmetry and to reflect on US policy regarding data dissemination. My principal thesis is that parties to arms-length minerals/petroleum negotiations have differential capacity to acquire, analyse and use Landsat data. These differentials are particularly significant when one negotiator is a multi-national minerals/petroleum exploration and extraction firm, and the other is a poor LDC. My arguments are related to the issues of minerals exploration and exploitation, although similar or analogous arguments could be made concerning international food commodity negotiations and speculation.

## Organisation of the US Satellite Programme

The Landsat programme is currently the responsibility of the National Oceanographic and Atmospheric Administration (NOAA) in the US Department of Commerce, although the National Aeronautic and Space Administration (NASA) retains responsibility for technology development and implementation in the near term. NASA will, for example, supervise the construction and launch of Landsat D and D′, scheduled respectively for launch in mid-1982 and mid-1983, and will be responsible for early command and control as well as for pre-operational testing and research on the high resolution, high data rate Thematic Mapper (TM) sensor which is to fly along with the Multi-Special Scanner – MSS) on D and D′. All of US earth sensing satellite activities are being centralised in NOAA, and the EROS Data Center, a part of the Department of the Interior, will operate as a contractor under financial and administrative control of NOAA and NASA will bow out entirely, it seems, after the TM is operational.

NOAA is building its own new, state-of-the-art image pre-processing facility and will take over this function as well as command and control from NASA in about January 1983. NOAA's role in remote sensing is seen as strictly transitional, to move the service from a quasi-experimental public service venture to a quasi-operational business-like operation to be taken over by private sector interests. The private sector operationalisation of Landsat (i.e. as a for-profit business) is only a question of time. There seems to be strong pressure from the Office of Management and Budget (OMB) to move it into the private sector as soon as possible.

NOAA is under a firm mandate to recover fully all *operating* costs associated with the Landsat programme. What is meant by operating costs had never been especially clearly defined, but they seem to include everything but new capital and debt service. These costs are on the order of $40,000,000 per year. There are three components to the revenue

**Table 7.1** *Data acquisition*

| *Source* | *Form* | *Resources Required* | *Time* |
|---|---|---|---|
| Direct from Satellite | Data Stream | $\$10^6 - \$10^7$ for a tracking/receiving station<br><br>Skilled cadre of engineers and technicians to operate and maintain station<br><br>Fees to NASA currently $200,000/year | Real time, no delay |
| EROS or Foreign Reception Station | Computer Compatible Tapes | Cost of reproduction: currently $200/tape | Weeks to months |
| EROS or Private Vendor | Enhanced Imagery | $1,000–$2,000 (or more) per scene | Months |
| EROS or Foreign Reception Station | Standard B/W or colour imagery (all *current* EROS products receive routine contrast stretch and haze removal | $8 for B/W<br>$50 for color composite | Weeks to months |

**Table 7.2** *Data analysis*

| *Technique(s)* | *Resources required* | *Outputs* |
|---|---|---|
| Pre-processing (geometric and radiometric correction, etc.) | High level data processing hardware $\$10^6$ | Computer compatible tapes |
| Digital classification<br>Statistical modelling<br>Image enhancement and Image restoration | Data processing facilities including (interactive) image processing hardware and software ($\$10^4$–$\$10^6$)<br>Geological and mathematical models<br>Scientists, engineers, programmers, statisticians, technicians/maintenance staff | Enhanced or restored imagery<br>Quantitative and qualitative inputs for decision models<br>Graphic information |
| Photointerpretation | Photointerpretation Lab $\$10^1$–$\$10^5$*<br>Skilled interpreters and technicians | Maps, graphic information<br>Qualitative inputs for decision models |
| Photointerpretation | Photointerpretation Lab $\$10^1$–$\$10^5$*<br>Skilled interpreters and technicians | Maps, graphic information<br>Qualitative inputs for decision models |

*A fully equipped Lab will cost $300,000 – $500,000
Simple equipment – tilt table and lens – cost $50

stream which would support operating costs. First, data sales to government agencies; second, data sales to private users through the EROS Data Centre; and third, foreign receiving site fees, (which are projected to contribute approximately $10,000,000 per year). The Administration is resolved to raise prices and to ensure private sector operationalisation.

## Access to satellite data

It is important to emphasise that there are enormous differences in the capabilities of using Landsat data in different developing countries.

**Table 7.3** *Information use*

| *Filter* | *Additional Inputs* | *Outputs* |
|---|---|---|
| | Mathematical decision models embodied in computer software. Integrated print, graphic and computer data bases. | Decisions |
| | Other remote sensing information (aeromagnetics, etc.). Ground experience/ground truth. | |
| Communications Infrastructure | Market information, experience | |
| | Print and graphic data bases. Other remote sensing information. Ground experience/ ground truth. | Decisions |
| Institutional Infrastructure which incorporates the output of data analysis into the decision-making process | Market information, experience | |
| | Print and graphic data bases. Other remote sensing information. Ground experience/ ground truth. | Decisions |
| | Market information, experience | |

Brazil[9] and India, both of which intend to launch their own remote sensing satellite systems by the end of the 1980s, have superior capability to most of the countries to which the following arguments are addressed.

In principle, the first step in using Landsat data is getting it, which is presented as a highly simplified model in Table 7.1. While Landsat data has until now been easily accessible, at least in some forms, it has not been unusual for countries which might make good use of it not to avail themselves of it. In Kenya, Lowe[10] noted that while Landsat imagery was being used rather successfully in some land use projects, no use was being made of it for petroleum exploration even though Kenya had the trained staff and the photo interpretation equipment, and a foreign multi-national petroleum company was exploring in Kenya. Benedict Levin, who was formerly with the Earth Satellite Corporation, recounted that one West African country engaged in various satellite remote sensing projects built a multi-million dollar deep water port in a silt zone, when a cursory examination of Landsat imagery would have warned the country to select another site. He also indicated that his Earthsat Corp. was called in to consult with a wealthy, technically advanced Middle Eastern country. Officials realised after completing a tough copper negotiation that the extraction firm, which had never even prospected the area on the ground, knew more about the country's copper resources from Landsat images than did the country's own negotiators.

Once a country decides to acquire data it may acquire it in a number of different forms. The least expensive data product is a black and white image made from 70 mm film and available from the US EROS Data Centre and foreign Landsat stations. For a good deal more money, perhaps $2,000, compared to $10.00 for a standard image, a computer-enhanced image may be purchased from EROS or a vendor who processes the computer compatible tapes. Prices for data products are rising sharply.[11]

The fastest and certainly most expensive way to acquire Landsat data is to build a receiving station to obtain telemetry directly from the satellite. The politics of Landsat receiving station construction is only tangentially related to the issue of differential utility of Landsat derived information, but it merits a brief digression. Since rather early in the Landsat programme the US government has encouraged the construction of foreign receiving sites, both as a backup to relatively unreliable on-board tape recorders, and as a way to encourage foreign governments to become more involved in the Landsat programme. Currently, there are two operating stations in Canada, one in Italy and one in Sweden (both operated under a memorandum of understanding with the European Space Agency). Other stations are in Brazil, Japan, Australia, India, Argentina and South Africa.[12]

In the next few years, foreign stations will be more important to the US than ever before. In technical terms, the current Landsat system is in deep difficulty. Landsat 2 is no longer operational, a problem not of electronics but of maintenance of spacecraft control. Landsat 3 is experiencing telemetry problems which lead to intermittent loss of data. Landsat D, launched July 1982, has no on-board data recorders. Since the Tracking Data Relay Satellite System (TDRSS) is not yet operational, four or more of the foreign receiving stations will be equipped with special data recorders to capture MSS data in real time, and tapes will be airmailed back to the US for pre-processing and formatting.

There is some controversy about the thematic mapper (TM): some NASA officials want it to be used strictly in a research mode with data distributed only to NASA-authorised investigators at least until the system is working extremely well. Others are pushing for public access as soon as any data is produced. The MSS and pre-processing facilities for D and D′ will be able to produce, with high levels of reliability. TM data is to be relayed via TDRSS. There are no current plans for foreign reception of TM data. This is important because TM data is potentially much richer than is current MSS data.

To acknowledge that foreign receiving sites are important for both US and foreign data acquisition is not to say that every foreign country should build its own Landsat receiving site. Even with the projected lower orbit of Landsat-D, a very few additional receiving stations will be required to provide quasi-total earth coverage. The US has encouraged a regional approach to foreign receiving stations by attempting to ensure that orders for data products go from a foreign country to the nearest foreign receiving site rather than back to the EROS Center in South Dakota. This is, in part, to help each foreign site support its operations through sales of data products (attempt has been made to standardise prices to discourage comparison shopping).

At first glance the regional strategy looks quite sound, but in Africa, for example, regional politics, national prestige, and US interests, having nothing whatever to do with remote sensing, may dictate a technically suboptimal placement of receivers resulting in redundant capacity in some areas and no data acquisition in others.[13] In South America, Argentina's recently completed station is more than 90 per cent redundant with Brazil's and, while it may be argued that the coverage the station provides of the off-shore islands and remote coastal areas is critical, it is safe to imagine that national prestige and not technical necessity had a great deal to do with the station's construction. Many of the nations currently considering constructing Landsat receivers would be creating redundant (it could be argued unnecessary) capacity. Far from improving the use of remote sensing for development, unnecessary duplication of coverage for reasons of national prestige or difficulties of regional co-operation may divert scarce resources – money, technology and skilled people – from the most essential activities: analysis, interpretation and use of the data.

Initially, the price of data products and annual receiving station fees were nominal, designed to encourage the use of Landsat data. Receiving site charges for foreign receiving stations currently pegged at $200,000 per year can be expected to go up by approximately a factor of 3 by 1984.[14] In addition, there is a suggestion that the foreign stations pay some sort of 'royalty' or 'distribution fee' to the US on the revenues they generate from data sales to third parties. These are sensitivc issues which are as yet unresolved.

Given the enormous cost of exploration and extraction of resources, the cost of Landsat data products, at ten (or perhaps even one-hundred) times their current price will be insignificant to the major multinational mineral and petroleum firms. It is fairly clear that they will be willing to pay whatever price is charged for the information. Apart from specially earmarked foreign aid (a possible prospect), some of the poorest of the LDCs, especially those with the lowest level of interpretive capacity, may choose not to purchase what may become for them very expensive data products.

*Data analysis*

Once the data is acquired from the satellite, it must be analysed and interpreted before it is of use. In the use of Landsat data for minerals or petroleum exploration, the interpreters must begin with an implicit or explicit geological model: 'A geological-geophysical description and characterisation (quantitative where possible) of the mineral being sought – its genesis, structural setting, composition, size, ore grade, etc.'.[15] Since exploration is a very important aspect of their business, the multinational mineral firms employ very sophisticated models as the basis of their Landsat analysis, which contrasts sharply with LDC capability. Certain skills and knowledge, which are often lacking in countries new to the technology, are required to perform even the simplest analysis from satellite data.

Photo-interpretation of low cost black and white data products is the least expensive, 'simplest' form of data analysis. Such analysis can be quite useful in minerals and oil exploration. Computer-enhanced images are available from EROS and private vendors and may be analysed using similar and more sophisticated methods of photo-interpretation. LDCs by no means always avail themselves of computer-enhanced images nor are they always staffed to be able optimally to interpret them.

Digital techniques are the most powerful methods of analysis for the interpretation of Landsat derived data. Digital analysis takes maximum advantage of the digital readout of the multi-spectral instrument, and may employ high powered mathematical/statistical models embodied in sophisticated computer software to draw rich inferences concerning mineral placement. Interactive image analysis, where a scientist or engineer at a computer terminal is able to 'message' the data in quasi-real time, has proven to be a particularly fruitful method of Landsat data interpretation. Multi-nation petroleum and mineral concerns have used and routinely use the highest levels of digital data analysis in conjunction with Landsat data; the LDCs do not.

Computer techniques are important analysis tools because when used properly (i.e. with an interpreter familiar with the area under study) they can extract information unavailable by manual techniques alone. Unfortunately, the transfer of digital image processing to LDCs is a very complex task. Digital image processing requires personnel trained in computer programming and electronics maintenance, as well as mathematicians with specialities in statistics and pattern recognition, if the digital systems are to stay in repair and remain adaptable to changes. These types of personnel are at best difficult to find and at worst unavailable in many LDCs. Computer technology is expensive: a fully equipped computer analysis laboratory represents a $800,000 to $1,000,000 investment. Most LDCs therefore would find use and maintenance of computer image processing costly and difficult to justify in their needs for development.[16]

The hardware for digital analysis is often not present in an LDC (see Table 7.2 for a schematic summary of the techniques and resources required for data analysis). But it is important to note that the cost of data processing hardware has been dropping sharply with the advent of large scale integration and miniaturisation. A good deal of software has been written or adapted for less costly systems; which in principal do anything one of the large systems would do, only slower.

A much more serious problem, one in no way amenable to a short term 'technical fix', is the lack of available trained manpower to do the job. The importance of a team approach has been continually stressed in the literature on satellite data analysis. In order to make optimum use of Landsat data, especially for minerals and petroleum exploration, a highly skilled multidisciplinary team of substantive experts – economic geologists, geophysicists, petroleum engineers, etc. – and high level support personnel – electrical engineers and mathematicians – as well as skilled technicians, programmers, electronics maintenance personnel, etc. is required.[17] Such skills are not readily available in the US, let alone in a poor LDC.

The response 'if there are no experts, train some' is facile and overlooks the reality and complexity of the training/technology transfer task. The US Agency for International Development and several other international development organisations support a number of fine remote sensing training centres in Latin America, Africa and Asia, but the courses are largely short-term, specialised programmes rather than long-term degree or post-graduate education. Moreover, in line with AID's primary mission in rural and agricultural development, the training at the AID-supported centres tends not to emphasise minerals exploration.[18]

Brain-drain has been a serious problem in many technology transfer efforts. The Report of the UN Conference on Science Technology and Development in Vienna in 1979 mentions the problem no fewer than six times. Brain-drain in remote sensing takes several forms. Scientists and engineers trained in foreign universities often simply never return to their country. The US major oil companies are suffering a brain-drain of their own[19] well trained and talented people. Those trained in quantitative analysis of remote sensing data are in particularly high demand. Some scientists trained at government expense for government jobs are lured away to private industry at home. Charles Paul of AID reported that MS and PhD level people trained in remote sensing related fields often 'disappear' from remote sensing activities, as they are promoted to administrative posts or become engaged in scientific research completely unrelated to remote sensing. The AID solution has been to: first, train people as close to home as possible, almost never bringing them to the US for training; second, train lower level people, technicians or others with the equivalent of one or two years of college; third, keep the training short and topical. The problem with these solutions is that they do not come to grips with a central problem: how to supply the remote sensing staffs of LDCs high level scientists of skills comparable with their counterparts in multinational corporations. Wolfram Drews of the World Bank suggested that the only way to combat brain-drain is to train three times as many people as you know that you will need.[20]

## *Information use*

Once the data has been acquired, analysed and interpreted, that is, once it has become information, an important step remains: to use it, by incorporating it into the process of planning and decision-making. One of the hallmarks of the modern multinational corporation is its ability to effectively handle and use information. The same cannot be said for most of the LDCs.

'Prospects' and technology transfer literature have consistently emphasised the importance of building institutional structures to foster the use of satellite derived information in developing countries. It seems that there has been significant progress in this area since the early and mid-seventies, but a great deal more remains to be done. Progress has been very rapid in some places and practically nonexistent in others. This is surely to be expected, since institution-building is a slow, even painstaking process. It is not uncommon that carefully analysed data and potentially useful information may be available but unused in a developing country, not because of a personal human error (which can and does happen everywhere, not least in large corporations), but because of inter-agency rivalry or a lack of organisational channels for the transmission and reception of the information. It is almost inconceivable that the information generated through the digital analysis of Landsat data would not be available to the negotiating team of a multinational minerals company. It is unfortunately quite conceivable that such information, even bought from an outside consulting firm, would not be available to the negotiators for the LDC.

'Using information' entails much more than incorporating a given type of data in a decision-making process; it involves combining many different kinds of information from a variety of sources. See Table 7.3 for a schematic presentation of this. One particularly important type of information, actually a locus of a great deal of varied information, might be characterised as ground experience.[21] An LDC has considerable potential bargaining advantage with respect to its superior ground experience and its ability to combine it with analysis of Landsat data. Indeed, it is often argued that this advantage is decisive, for example, more than outweighing the much greater analytical capabilities of the multinational firm. Whether this is true or not is an empirical question which can only be answered in relation to the analysis a concrete factual situation.[22] As a general proposition, I find it unlikely. Certainly, it does not jibe with what I have learned about the use of Landsat in specific minerals negotiations or with the usual situation in such negotiations.

Following a trend in the field of information management, sophisticated users of satellite information are developing integrated information systems or data bases in which Landsat data, in digital form, may be combined with other computer data to be stored and accessed in machine readable form. In a recent paper[23] on the Integrated Data System Concept, it was noted that most research and development efforts have been devoted to 'front end' issues, including scanner design, or reception systems, and that considerably less attention has been paid to the user models (for information analysis) which converts the data to useful knowledge. Another work on the subject holds that 'both the temporal and spatial frequency of the data – not to mention the format of the data – may be beyond the capacity of the models of the information systems that support them'.[24] The paper argues that technology transfer efforts in this realm have been especially inadequate. The startling fact is that they are referring to technology transfer to US state and local governments, relatively sophisticated computer users with large budgets, trained staffs and ready access to consultants from industry, government, universities and research institutes![25]

A US government report on the subject made the following assessment:

> Very few developing countries currently have the facilities, trained personnel, or institutional structures to make good use of the data available from the Landsat. Indeed, the level of satellite technology is such that it will require many years and much investment before developing countries generally will be able to systematically apply data to their development needs.[26]

Although the US open dissemination policy assures all developing countries access to data, many do not avail themselves of it. Sometimes issues of prestige or policy may hamper data acquisition or may dictate unnecessary expenditure for high cost, low utility prestige items like excess receiving stations. Price increases will continue to limit dissemination of data products in the poorest of the LDCs. When a country does obtain data, it often does so in the simplest, least expensive and generally least valuable form. Multinational mineral and petroleum companies regularly obtain data directly from the US Government dissemination centre or in an enhanced form from specialised vendors who service the industry. They normally obtain higher cost, higher quality data products including computer compatible tapes.

Multinational corporations have vastly superior capability for the analysis and interpretation of Landsat data. In particular, their ability to perform digital analyses gives them more valuable information than an LDC is able to obtain through visual techniques alone. Facility problems which have traditionally plagued the LDCs, may be partially alleviated by the reduction in computing cost; they are not so acute as the lack of trained personnel. Given the long lead time required to train a skilled quantitative analyst along with the problems of brain-drain and the high demand for technical personnel in LDCs, it is unlikely that the problem will be fully solved in the foreseeable future.[27]

Even with the best information provided by the most honest, competent foreign consultant, expatriate expert or national resources analysis department, an LDC is likely to be at a disadvantage in its capacity to use the information effectively in negotiation. Local ground experience, even if it is incorporated in the negotiating process, is often more than outweighed by the multinational's access to print, graphic and computer data bases which may include, in addition to the best available analyses of Landsat data, highly sophisticated geological models, information about world supplies, market projections, contracts and negotiations in other countries and high level decision-software for handling all the data. The net result is that in some cases Landsat data has redounded to the detriment of the poor LDC in minerals and petroleum negotiations. This is in no way to gainsay the positive value of Landsat data for LDCs in other areas. Bolivia and Peru, for example, have been signally successful in striking advantageous bargains in no small measure due to their knowledge acquired through Landsat. A much less happy and perhaps more frequent scenario is the following:

> A negotiation is undertaken. The multinational has never explored Country X but Landsat information points to a strong possibility of copper in Area A and a weak possibility in Area B. The company makes Area B the focus of negotiation and gets a concession at Z price. In passing they also get a concession at one-half Z in Area A which 'doesn't look very promising and besides will be very expensive to exploit'. A major find is made in Area A. Country X becomes suspicious and wants to renegotiate the deal *ex post facto*. The company argues that a contract is a contract. Country X expropriates and nationalises the field. Without adequate capital or technology, mining operations grind to a halt. Everybody loses.

In the introductory chapter to this book, Cruise O'Brien and Helleiner offer another, perhaps more typical, scenario where the company does not try to deceive the host country but merely begins with more and better information. To the extent that differential usability of Landsat data contributes either to deception (not entirely unknown) or merely to information imbalance, it may be in the long-term interest of neither the resource company nor the LDC.

## Conclusion

'Data' may have no intrinsic value; 'information' does. Differential capacities to interpret data and to use information may be directly related to differential strength in bilateral bargaining. If such imbalance has developed since the beginning of the Landsat programme, it is likely to be exacerbated by the new generation of high resolution sensors which will fly on Landsat-D (1982) and subsequent satellites. In spite of all of the efforts at training and technology transfer, it is by no means clear that the capacity gap between the poorer LDCs and the high technology minerals and petroleum multinationals is not widening.

The official US position, as well as that of many of its apologists[28] is to my mind quite simply wrong when it asserts that an LDC's territorial control over resources 'on the ground' and equal access to Landsat data are more than sufficient means to protect its interests in arms length negotiations. To put it another way, a crucial question for developing countries is how power over resources may be converted into power over events. It does not require a great deal of imagination to recognise that (at least in some cases) open access to Landsat data may be part of the problem rather than part of the solution.

A possible response to the present impasse in international political circles might be to restrict third party access to Landsat data by permitting the sensed nation to decide who may have it and who may not (the Franco-Soviet proposal). In the introduction it was noted that US policy on access has been remarkably consistent over the years and administration; but from at least one point of view, it has been consistently schizophrenic: 'civilian' remote sensing data is entirely open while 'military' data is entirely secret.[29] It has been difficult to ascertain the precise US posture on Seasat type data (to be generated by the National Oceanic Satellite System), but it is clear that the Department of Commerce is unlikely to provide Peru, for example, real-time reception of data allowing her to pinpoint the US tuna fleet, and the Navy may be loathe to provide China with direct access to information on the precise location of its ships at sea. When NOSS data is made available, I think it is fairly safe to assume that 'acess' will be liberally interpreted not to compromise US national and economic security.

In the (extremely unlikely) event that the US were to change its policy on Landsat data – that is, to bring it in line with international opinion and its own policy on data from other satellite systems – it might be argued that insuperable technical problems would make restriction impossible. First, Landsat cannot 'see' political boundaries – the data is read out by scene, not by country. Second, 'the cat is already out of the bag' – the multinational mineral and petroleum firms already have much of the data in hand from eight years of Landsat operations. Besides, even if new data were restricted it would be nearly impossible to enforce the restriction. These are serious problems, but as they are basically technical they may lend themselves to technical solutions.

It does not, for example, seem inconceivable that a mask in the shape of a country's boundaries might be electronically overlayed on a scene. The data within the boundaries could then be separated out of the total scene and rerecorded for dissemination. This would add an extra step to the task of data 'preprocessing' and it might be expensive, but some such system is probably not beyond the capacities of the computer scientists. Further, although there has been nearly a decade of sensing already, even more valuable data is several years away; restriction could begin with the advent of the thematic mapper (TM) and other advanced sensors, such as the multi-linear array, the new French satellite will carry. Short of military classification, it will be difficult to eliminate entirely unauthorised possession of TM data, but if such possession is indeed unauthorised, then possession would be prima-facie evidence of bad faith.

A further technical objection to access restriction is that one of the main values of Landsat data is its intrinsically international scope. Many applications of Landsat in agriculture, urban and land use planning, forestry, are national in nature and of little interest to third parties; data restriction would have no impact on such uses. Nevertheless, there are other very important applications, especially in environmental monitoring of pollution, the development of deserts, snow or water conditions, etc. which are regional or international and where access restriction could indeed vitiate the value of the data.[30] It may be possible through regional organisations to provide data to concerned parties (nations) without releasing it to multinational companies but, as noted above, regional organisations may look better on paper than they are in practice.

I have not seen an adequate mechanism suggested at once to limit access by requiring the consent of the sensed country while providing access to protect the legitimate interests of other countries and the international scientific community. Even if, for the sake of argument, we grant that a sensed country does have the right to protect information about its mineral resources, such restrictions would have serious negative consequences having nothing to do with mineral resources *per se*. It will be interesting to see how the French deal with the issue of data access and restriction given their historical position in the UN debates.

Several of the legal works as well as US government policy statements have highlighted the centrality of the Freedom of Information Act in barring any effective restrictions on data dissemination. The argument goes that the Freedom of Information Act requires that unclassified information collected by the Federal Government, including Landsat data, must be available to interested US citizens. Thus, while the US government might wish to agree to restrict data dissemination, it could not do so. This is an intriguing argument, but one that is probably beside the point. The Commerce Department transition document, *Planning for a Civil Operational Land Remote Sensing Satellite System* (June, 1980), suggests that changes will have to be made in the Freedom of Information Act with respect to Landsat data, in order to protect the profitability of an eventual private sector system operator. If the FOIA could be amended to protect American private enterprise, there would seem to be no legal barriers to changing it in accordance with a new international treaty.

Developing technical and legal mechanisms which would protect a nation's legitimate interests without doing violence to its neighbours or the international community's interests would be strictly academic, since the US domestic situation is almost certain to preclude any change in its longstanding open-dissemination policy. The reason access will not be restricted is that there are important US economic/political interests,[31] as well as a tradition of free flow of information,[32] which I believe will block any initiatives to restrict access to Landsat data. The largest non-governmental users of Landsat data are the petroleum and minerals industries. Certainly, until now, the most important direct economic benefits from Landsat have been in the area of economic geology. In the future the highest money stakes in remote sensing may, however, turn out to be in agricultural commodities rather than minerals.

Since 1976, the US-industry sponsored Geosat Committee, an extremely effective and articulate organisation, has presented the interests of the exploration geology community to the government and made significant contributions to the advancement of the Landsat programme.[33] The Geosat position is that resources are 'international':[34] one would be hard pressed to find a leader of a resource rich, capital/technology poor LDC, who agreed with this position. It is by no means frivolous to hold, as does Geosat, that open access is likely to lead to more discoveries more quickly than restrictive access. Whether fast discoveries which a sensed country might resent will lead to more production in the long run is an open question.

A complete *laissez-faire* approach, which is advocated by some in the Reagan Administration, could lead to the discontinuance of US satellite remote sensing operations after Landsat D′ or to a completely private proprietary operation. Since capital investment for an operational system is in the $600 million to $1 billion range, finding a private sector operator will not be easy unless it is guaranteed some substantial government support – special tax write-offs, direct subsidies, etc. If an operator could maintain exclusive use of the data, it would be quite easy to raise the $1 billion required because that much could be made by getting an advanced forecast of only one major global crop failure. Since there is a 90 per cent probability of three such failures within a decade, it is not difficult to imagine the incentive to get exclusive access to this data. Exclusive access, though, seems extremely unlikely and so we are faced with an almost classic public goods problem – the kind of problem in which the current Administration seems not particularly interested or able to solve.

If we accept, given the current US political climate, that restrictions of data access are impossible, it is still possible to chip away at the negotiation problem. One small but symbolically (and perhaps eventually even substantively) significant step would be to inform the sensed nation of requests for images of its territory. This would require on additional form to be filled out and mailed for each scene of a foreign country requested. The sensed country would then, at a minimum, know who was interested in its territory. This might mitigate the information imbalance by allowing the sensed nation to plan its own analysis of Landsat or ground data well before formal negotiations begin. But given the arms-length, competitive nature of future negotiations and the probable continuity of US policy on open data dissemination, along with an effort to make the National Environmental Satellite Service self-supporting from user fees, I think the issue of asymmetrical utility will be with us for a long time.

## Notes

1 Ronald F. Stowe, 'Diplomatic and Legal Aspects of Remote Sensing'. Paper presented at the 41st Annual Meeting American Society of Photogrammetry (March 9–14, 1975).

2 Celso Antonio DeSouza E. Silva, 'Treaty on Remote Sensing of Natural Resources by Satellite – Draft Basic Articles', UN Document A/AC. 105/122 (Brazilian Mission to the United Nations, 1974).

3 Leonard Jaffe, Press Release of Statement by US Representative to the United Nations Committee on the Peaceful Uses of Outer Space Combined Session of the Working Group on Remote Sensing of the Earth by Satellite and the Task Force of the Working Group (US Mission to the United Nations, February 25, 1974).
I. A. Vlasic, 'The Evolution of the International Code of Conduct to Govern Remote Sensing by Satellite: Progress Report,' *Annals of Air and Space Law*, Vol. 3 (1978).

4 Gary L. Hopkins, 'Legal Implications of Remote Sensing of Earth Resources by Satellite', *Military Law Review*, Vol. 78 (Department of the Army, Washington, DC, Fall 1977).
Hamilton DeSaussure, 'Remote Sensing by Satellite: What Future for an International Regime?', *The American Journal of International Law*, Vol. 71 (1977).
J. J. Hahn, 'Developments Toward a Regime for Control of Remote Sensing from Outer Space', *Journal of International Law & Economics*, Vol. 12, No. 3 (1978).
Nicolas M. Matte and Hamilton DeSaussure (Eds), *Legal Implications of Remote Sensing from Outer Space*, (Sijthoff, Leyden, 1976).

Gregory E. Moore, 'Earth Resource Satellites, A Puzzle for the United Nations', *Harvard International Law Journal*, Vol. 16, No. 3 (1975).

5 Nelson J. Parada, 'Collection, Processing and Distribution of Remote Sensing Data from the Brazilian Receiving Station', 14th International Symposium on Remote Sensing of the Environment (ISRSE), (1980).

6 National Academy of Sciences, *Resource Sensing from Space: Prospects for Developing Countries* (NAS, Washington, DC, 1977), p. 147.

7 Ibid. pp. 145–149 and
Valerie Hood, *A Global Satellite Observation System for Earth Resources: Problems and Prospects*, Report to the National Science Foundation (American Society of International Law, Washington, DC, n.d.), pp. 125–129.

8 DeSaussure, op. cit.

9 The Brazilian Institute for Space Research, for example, (INPE) has a remote sensory research staff of over 100. The image production department alone has a staff of 80 (20 per cent having a BS or higher degree) and an annual operating budget of $1.2 million.

10 D. S. Lowe, *et al.*, *An Economic Evaluation of the Utility of ERTS Data for Developing Countries* (Environmental Research Institute of Michigan, 1974).

11 EROS has about doubled prices in the last year and we can expect them to go up about 2.5 times the current level in October 1982 (that is 3.5 times last year's prices). In addition to the new price schedule for archival/retrospective data, a new 'special acquisition' price structure will be put in place. For a premium rate of about $1,200 a user will be able to *request* a specific scene as is.

12 Memoranda of understanding have been signed with China (station to be operational in 1981 or 1982) and Thailand (currently procuring hardware). Zaire and Chile were authorised to build stations but their memoranda of understanding have lapsed and no construction ever took place. New Zealand, Venezuela, Ecuador, Nigeria, Korea, Indonesia, the Philippines, Pakistan, and Romania, among others, have expressed an interest in constructing Landsat reception facilities.

13 The African Remote Sensing Council may be able to overcome some of these problems eventually (cf. Ouedraogo, 1980 for an account of the Council's development and accomplishments to date). At this point the status of regional cooperation in remote sensing in Africa remains unclear.

14 It is not unreasonable to think that prices will be pegged at what the exploration firms are willing to pay since they are the largest non-governmental user group and the group which has realised the most concrete pay-off from the data.

15 J. Bale, *et al.* (Eds), *Remote Sensing Applications to Resource Management Problems in the Sahel*, Chapter 6 'Geology & Mineral Prospecting,' M. C. Erskine and S. B. Levin (Earth Satellite Corporation, July 1974, prepared for AID, 1974).

16 C. K. Paul, 'Satellites and World Food Resources', *Technology Review*, Vol. 82, No. 1 (October 1979), p. 27.

17 Some petroleum companies use a headquarters-based, intensive team approach to Landsat analysis, while others train their field staff to use the imagery *in situ*.

18 William C. Draeger, 'Opportunities in Remote Sensing Training for the International Scientific Community', 14th *ISRSE* (1980) provides a list of training programmes.

19 cf. 'Manhunt in the Oil Fields', by A. Stuart, *Fortune*, October 6, 1980. Exploration geologists are being wooed from the major oil companies by smaller, higher paying exploration companies, *not* by the resource ministries of poor developing countries. This creates an even greater incentive for brain-drain from the LDCs to the major oil companies.

20 F. P. Dubois and W. D. Bruce, 'Remote Sensing Technology Transfer the Canadian-Peruvian Approach', *13th ISRSE* (1979).
Guy Rochon, *et al.*, 'An Experimental Technology Transfer Project in Remote Sensing between Canada and West Africa', *14th ISRSE* (1980).

21 By ground experience I mean a great deal more than the technical concept of 'ground truth' which normally refers to information or data to calibrate, validate or confirm remote sensing information. By ground experience I understand a rich, intimate familiarity with all aspects of a given territory.

22 See C. Fortin, Junguito and Pizano, *q.v.* on commodities.

23 David W. Keller, & George G. Frippel, 'AIDS: An Advanced Integrated Data System Concept', *13th ISRSE* (1979).

24 Ronald L. Shelton and John E. Estes, 'Integration of Remote Sensing and Geographic Information Systems', *13th ISRSE* (1979).

25 An employee of the Chevron oil field research company suggested that integrated systems may be more developed and important in land use and urban planning than in minerals exploration. This not entirely borne out by the literature. The importance of data base integration in minerals exploration is not entirely clear.

26 'A Preliminary Assessment of AID's Requirements for a Landsat Follow on Satellite'.

27 Some countries rely on expatriate experts, a short-term solution which may increase a country's long term dependency. Some LDCs are loathe to depend on consulting firms from the North as a few have been (or perceive themselves to have been) 'burned' by unscrupulous operators who have kept the most valuable information for themselves. Trained indigenous personnel is the only long-term solution.

28 See, in particular G. L. Hopkins, 1977. Op. cit.

29 As far as I know, declassification of economically valuable data from military satellites is still in the talking stages. Cf. Geosat 1979 and US Senate 1979.

30 A subjacent country might have many more or less legitimate reasons for wanting to limit access to Landsat data: military, political, etc. One can imagine that an upstream country which is polluting a river might not want its downstream neighbours to have access to evidence of its malfeasance. I would suggest that this represents a rather less legitimate class of reasons to restrict data but probably not an altogether uncommon one.

31 Vlasic (1978), op. cit., in an article on the evolution of the international code of conduct in remote sensing, wrote that '. . . once certain practices become established and vested interests develop, it will be well-nigh impossible to obtain an [international] agreement which would have the effect of altering the status quo' (p. 571).

32 Perhaps 'mythology' would be a more apt term because the tradition

tends to crumble in the face of real national security or commercial interests.

33 For example, that there will be a 2.2 micron band sensor on the thematic mapper is largely due to the exertions of the Geosat Committee. This addition is a significant improvement in the instrument. Some observers have gone so far as to suggest that were it not for the industries represented by Geosat, there would be no ongoing US remote sensing programme. A recent development is a plan to use imagining provided by the space shuttle for the identification of mineral deposits.

34 This is justified in the Panel Discussion, 'Private Sector Involvement in Remote Sensing', Moderator: C. W. Mathews; Participants: F. B. Henderson-Geosat Committee, L. L. Farnham-G. E., A. K. Parker-Daedalus Enterprises, R. E. Tokerud-Lockheed Electronics, D. S. Lowe-ERIM, *13th ISRSE* (1979).

# 8

# Intelligence Requirements and Information Management for Developing Countries

Nicolas Jéquier

Judging solely from an overview of the economic and business literature, one can probably state that three of the most important preoccupations of the 1970s were (a) energy, (b) money, and (c) transfer of technology. One of the dominant themes of the 1980s will almost certainly be information, and what I would like to do here is not simply to join the new intellectual bandwagon, but explore some of the more practical and down-to-earth ways in which developing countries could exploit information in a more effective way.

Before doing so, however, it may be useful to state some of the main assumptions, hypotheses and preconceived ideas underlying this paper. These may be briefly summarised as follows:

(i) Knowledge in general, and information in particular, is an important yet dramatically underexploited resource in the process of economic and social development;

(ii) The mobilisation of knowledge and information for development purposes, if it is to be successful, will depend essentially on the internal organisational capabilities of the countries concerned, and much less than generally believed on concerted international actions;

(iii) Technological advances in the field of hardware are opening up major new opportunities for developing countries, but the fascination with hardware tends to obscure the need for software and seems to be leading to a dramatic neglect of the more simple and less costly ways of building-up an information-efficient infrastructure;

(iv) The sophistication of a country's information infrastructure is closely correlated with national wealth (rich countries are also information-rich countries), and, while the evidence for a causal relationship between information-intensity and level of development may be tenuous, there are good reasons to believe that higher investments in information can have more than proportional effects on economic growth and social development.

## A few basic concepts

One of the difficulties in any analysis of the information problems and challenges facing developing countries is that of terminology. For practical purposes, it may be useful here to make a basic distinction between information *levels* and information *functions*.

Information levels can be schematically represented in the form of a simple pyramid:

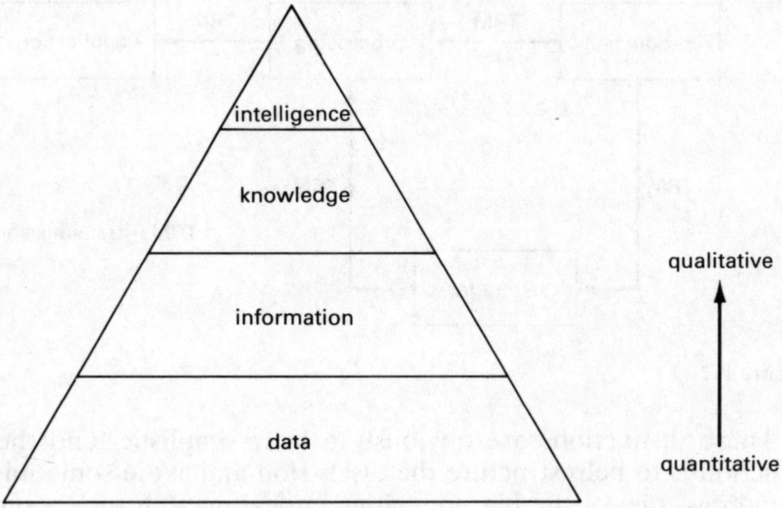


**Figure 8.1**

The frontier between these levels is one of degree rather than of nature: a computer print-out of seismic data can convey important information to the oil-exploration specialist, but remains a set of meaningless data for the layman. In the same way, economic information in the form of statistical charts and tables can add a lot to the knowledge of the specialist, but will remain irrelevant to almost everyone else.

Without trying here to give a standardised definition of the four concepts presented in this pyramid, two points of clarification may be useful. The first is that the term 'information' can be understood to mean structured and meaningful data (this is the narrow information theory definition of the word) or else can be viewed as the sum of all the information levels presented in the pyramid (this is the wider and more general definition which, unless otherwise specified, will be used here).

The second point to keep in mind is that as one moves from the bottom to the top of the pyramid, information becomes increasingly more qualitative and involves a higher proportion of complex judgemental and evaluative processes.

As for information functions, one can single out five basic categories – production, storage, processing, application and transmission – which may be schematically represented in a simple flow chart of the following type. Each of these five basic functions can be found at the various levels identified above: data are stored and processed in a computer and transmitted via a telephone line, while knowledge can be stored in a book or a document, and processed by the human brain.

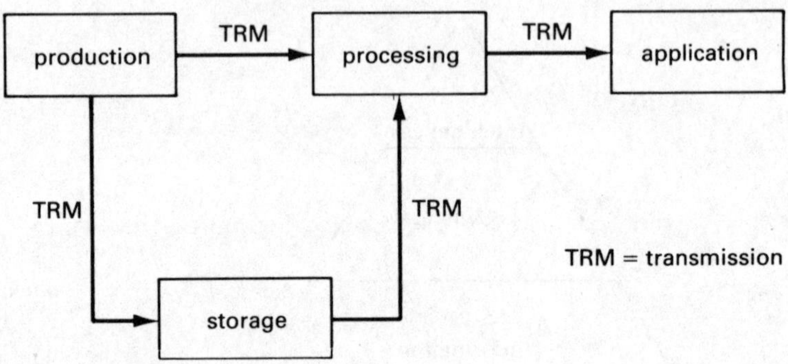


**Figure 8.2**

These distinctions are obviously a little simplistic, and their only function is to help structure the discussion and avoid some misunderstandings: one of the big difficulties in dealing with such a subject as information access and international negotiations is that the most basic terms often mean different things to different people. What is more, the focus of discussion is seldom the same: for one author, information access may mean primarily the access to data (how, at what cost and under what conditions can a developing country gain access to a foreign data base?), while for another, the problem is essentially to acquire certain pieces of knowledge which are important for development, or more specifically for international negotiations. Similarly, for one author, the critically important issue might be the hardware dimensions of information systems (computers, telecommunications networks, etc.), while for another the central point might be the problem of developing a knowledge industry and mastering the social problems of intelligence (how for instance does one develop the capability to make judgements about technological trends or improve a society's capacity to apply and act upon available information?).

## The problem

Most of the available evidence shows that the developing countries taken as a group have an underdeveloped information infrastructure: with a total population of 3.5 billion people (78 per cent of the world total), they have, for instance, only 7 per cent of the world's telephones, and a slightly lower proportion of the world's computers. Their share in the production of knowledge – 3 per cent of the world's research and development expenditures and around 5 per cent of its scientific and technical publications – is well below their share of industrial production (around 11 per cent) or their share in world trade (20 per cent).

Figures such as these may be useful in pointing to a number of important global issues, but their practical value is somewhat limited. There are a number of reasons for this. First is of course their macroeconomic nature: figures which aggregate countries as diverse as Singapore and Bangladesh, or Burundi and Brazil are always a little misleading, to say the least. Second is the fact that whatever may be done by the developing countries, individually or collectively, these global imbalances in the production and use of information between rich and poor nations are likely to persist at least until the end of the present century, if not far beyond. Third – and this is perhaps the most important – aggregate figures of this nature tell us little about what should be done, save for suggesting indirectly that the primary goal of the developing nations should be to catch up with their more wealthy partners in the Northern hemisphere.

What I would like to suggest here is that we try for one moment to forget these global data and look at some of the ways – cheap or expensive, trivial or important – in which developing countries can gain better access to the world's stock of knowledge, participate more actively in the international knowledge industry, acquire the information they need for development, and build up their own capacity to produce and use information. My underlying theme, to which I shall return at the end of this paper, is the need for organised information, and more specifically for what might be called an intelligence capability for development. The first sector, or field of action, I would like to cover here, is that of telecommunications, which happens to be one of the most expensive and capital intensive types of infrastructural investment a country can make, and which, partly for this reason, tends to receive far less attention than it deserves.

### *Telecommunications infrastructures*

Telephones are expensive, they are used primarily by the more affluent urban population, and national telecommunications authorities, which in many countries also operate the postal services, are big consumers of scarce capital and are often perennial loss-makers. In the political competition for funding, the telephone fares poorly, and public investment tends to go in priority to such obviously important sectors as agriculture, fertiliser production, education, health, power generation and of course defence. This phenomenon, it may be noted, is not typical of the developing countries alone: witness the case of France, which, until the end of the 1970s, was by far the least developed of the industrialised countries as far as telecommunications are concerned.

What is interesting to observe is that this tendency of most governments to underinvest in telecommunications[1] is to a large extent reinforced by the actions of aid agencies and international development banks: with the new emphasis given by these institutions to the fight against poverty, the satisfaction of basic needs and the promotion of rural development, the telephone appears even more irrelevant to development, and this accounts in no small part for the dramatic decline in the last ten years of telecommunications projects in foreign aid programmes and in the portfolio of investments of international development banks.

To state that telecommunications are a critically important instrument of development is essentially a matter of faith, but faith is not a sufficiently strong argument to convince finance ministries, planning commissions, councils of ministers, trade-unions or political parties, and weighs very little in the calculations of the economist. What are needed are facts and especially data to support the faith. Until quite recently, there were very few data to support the view that investments in telecommunication can be extraordinarily profitable, both from a social and economic point of view, and the usual argument for deciding for or against increased investments in telecommunications was the conventional financial one. On this score, telecommunications do not fare particularly well: in countries with a reasonably well-run telephone system, the net return on investment is in the 5 to 15 per cent range, i.e. well below the majority of conventional development projects.

The problem in essence is to try to identify and especially measure the indirect benefits of investments in telecommunications, and to carry out the same sort of analytical work in this sector as has been done, for instance, in the case of education. A series of studies on the indirect benefits of investments in telecommunications recently carried out under the sponsorship of the International Telecommunications Union (ITU) and the Development Centre of the Organisation for Economic Co-operation and Development (OECD) suggests that such investments are indeed extremely profitable from a social and economic point of view, and make a much greater contribution to economic development than was generally believed.[2] Without going here into the details of these studies, it may be useful to summarise some of their main highlights.

In the United States for instance, it has been shown that the financial support given by the Rural Electrification Administration (REA) to rural telephones in the form of interest subsidies has contributed very substantially to increases in rural income: according to E. B. Parker's figures,[3] every $100 invested by the Federal Government in this programme has contributed to a direct increase in gross national product of between $600 and $700, and to an increase in tax revenues which exceeds the direct cost of the programme to the government. A somewhat similar but much wider study carried out by A. Hardy suggests that between 1960 and 1973, an increase of one telephone per hundred people contributes to 0.01 per cent increase in gross domestic product (GDP) the following year in regions or countries with a high telephone density (80 telephones per 100 inhabitants), but that this contribution to GDP is approximately nine times higher (0.09 per cent of GDP) in low density areas (20 telephones per 100 inhabitants).[4] In other terms, the lower the density of the telecommunications network, the greater the contributions of telecommunications investments to increases in national income.

A case study of rural telephones in Egypt points in the same direction. According to A. A. Kamal and his associates, the average net benefit of the telephone to its user is several times higher than the monthly subscription charge.[5] For the handicrafts sector, for instance, the net measurable benefit of a telephone is approximately 80 times higher than the subscription charge. For users of capital equipment (i.e. small and medium size industries in rural areas), the benefits are even higher: over 100 times the subscription charge. This study, which is based on a detailed analysis of 146 villages, also suggests that the benefits of telephone service are directly correlated with a village's distance from the capital and the closest major urban centre: the more distant the village, the greater the net benefit. What is interesting to note is that these benefits, which turned out to be much higher than anticipated, are achieved with a telephone system of notoriously poor quality. They would certainly be even higher with a more reliable network.

A somewhat similar study carried out by M. Tyler in Kenya[6] tried to measure the losses sustained by a sample of nine typical firms as a result of poor telecommunications services, and to compare these losses with the cost of providing an adequate service. The evidence, here again, is compelling: the gains from better telecommunications service are on the average 100 times higher than the cost of providing this better service.

Other studies carried out in the framework of this ITU-OECD projet – for instance on telecommunications and public health services, on the export of agricultural crops, or on the global losses accruing to a national economy as a result of poor telecommunications – all point in the same direction: telecommunications are a major contributor to economic development, and the indirect benefits from such investments are several orders of magnitude higher than the direct benefits.

In addition to their obvious economic benefits, telecommunications are one of the most important infrastructures in the production, transmission and utilisation of information and knowledge, and one of the main preconditions for building up an effective national capability in the information field. Telecommunications are not cheap: given an average cost of $1,000 per telephone line, the total investments required to bring a country's telecommunications infrastructure up to international standards are very substantial. However, the developing countries are faced today with a set of economic and technological conditions which are much more favourable than those faced by the industrialised countries in the last hundred years.

First is the fact that the international market for telecommunications equipment is one of the most open in the world, contrary to the internal telecommunications markets of the industrialised countries, which are dominated by cartel arrangements between the main producers and important non-tariff barriers to the importation of foreign equipment. As a result of the fierce competition between the big international suppliers of telecommunications equipment, developing countries can build up their infrastructure at a cost which is significantly lower than that paid by industrialised countries. What is more, they can benefit from export credits, which at today's international interest rates amount to very significant subsidies from the equipment exporting countries.

Second is the advantage of the latecomer: in many developing countries, the building-up of a telephone system can start almost from scratch, and is unencumbered by the technical inefficiencies of obsolete equipment and the weight of earlier technological mistakes.

Third is the rapid decrease in the cost of hardware and its increasing efficiency, brought about by major technological advances in design and manufacturing. With fully electronic digital telephone exchanges, for instance, annual maintenance costs can be substantially reduced, and the quality of the network significantly enhanced. With small solar-powered earth receiving stations, it is now technically possible and economically viable to install telephones in the most remote rural areas and, what is equally important, to make the cost of transmission total independent of the physical distance between callers.

Fourth – and this is important from a financial point of view – it is organisationally possible (although not always politically feasible) to develop the telecommunications network on a self-financing basis. In all developing countries, there is a tremendous pent-up demand for telecommunications services, which is only very indirectly reflected in the length of the waiting lists for a telephone. Indicative of this demand is the black market rate for a telephone in the capitals of many developing countries (up to four or five times the average $1,000 cost of a telephone) or the colossal increases in telephone traffic once adequate service is eventually provided. Indicative of this is for instance the ten-fold increase in calls between Tehran and Isfahan in the month that followed the installation of an adequate microwave link between the two cities in the early 1970s, and the similar size increase in international calls from Oman following the installation of a satellite ground station. This pent-up demand, which in many countries far exceeds the total number of telephones now in service, also means that there is a willingness of potential subscribers to pay for what, to them, is a vitally important service. What is more, with some creative financing procedures (as the Japanese have been using since the 1950s and 1960s, with their organised official black market for telephones), it is possible to pay for the building up of a telecommunications infrastructure without any direct drain on the public exchequer.

### *The paper industry*

One of the main means for storing and transmitting information is paper, but the advent of electronics has tended to relegate this medium into the background, and very little attention has been given by information specialists to the effects upon the circulation of information of an underdeveloped or inadequate paper industry. In the industrialised countries, we take it almost for granted that ample supplies of inexpensive, good quality paper are always available for printing books and magazines, photocopying documents or packaging industrial products and processed food. In many poor countries, this is simply not the case, and the negative multiplier effects of poor quality paper and structural shortages of paper are enormous.

Basically, the problem can be viewed as going back to forestry, which provides the raw material for making paper. Differences at this stage between industrialised and developing countries can be significant. Witness for instance the fact that with a population and forest resources approximately equivalent to those of Finland but with much more favorable climatic conditions, a country such the Congo exports sixty times less paper products than Finland. And in India, the average wood yield per hectare is a bare 0.3 cubic metres per year, as against an average of over 3 cubic metres in industrialised countries which practice an extensive type of forestry (without fertilisers, and relying entirely on natural growth) and between 10 and 20 cubic metres for intensive types of forestry.[7]

The systems effects of paper shortages can have a number of surprising and unexpected effects. One case in point is the public health hazard brought about in India by the use of jute bags for cement: because of the shortage of paper and the unavailability of high quality kraft paper, the Indian cement industry stores all its cement in jute bags, and the cement dust polluting the atmosphere of construction sites because of the massive seepages of cement from the loosely knit jute bags is poisoning not only the construction workers, but also their children which live with them on the construction site. As for the losses which occur as a result of seepage during transportation or because of humidity at the storage point, they probably account for over 20 per cent of India's total cement production.

The effects of paper shortages and poor quality paper on the circulation and storage of information are fairly easy to document, but more difficult to measure. One may simply reflect here upon the fact that books printed on poor paper tend to self-destruct after a few years of shelf-life in public and private libraries, or upon the effects on the scientific community of a structural shortage of locally-produced journals, or the unavailability of good paper to use in the few and often mal-functioning xerox machines. In the same way, it may be interesting to reflect upon the difficulties experienced by the school system (an essential institution in the transmission of knowledge and information) in several of the poorest Black African countries as a result of the unavailability of textbooks and notebooks.

Historical analogies may be a useful pointer to the types of analyses which could be carried out on the relationship between paper and the diffusion, storage and production of knowledge and information. We know for instance that one of the reasons for the rapid development of printing in the 15th century Europe was the concomitant development of paper made from rags, which replaced the immensely more expensive vellum sheets used for manuscripts. This decrease in the cost of information storage was of at least one order of magnitude, and contributed in turn to a massive increase in demand for literacy and education.

A similar revolution occurred in the 1820s, with the invention of a process for manufacturing paper from wood. Cost decreases were at least of the same order, and there is an obvious linkage between the dramatic fall in the cost of paper between 1820 and 1850 and the expansion of the first mass media (the newspaper) and the development of primary education immediately thereafter.

Developing a national paper industry is one of the obviously important if apparently trivial steps in developing a country's ability to store and transmit information, and what I would like to suggest here is that a study be undertaken on the linkages between paper production and the development of information, to help identify the costs and benefits of investment in this sector, and make a case for the promotion of the paper industry. The underlying assumption here is that many countries have dramatically neglected their paper industry, and that this is having a number of negative incidences on the efficient circulation of information. Investments in the paper industry are probably as important as investments in telecommunications, computers or education, and have a major part to play in any government policy for the development of information.

## The fluidity of scientific and technological communications

Good telecommunications facilities and adequate supplies of paper are two of the important material elements of an information-efficient infrastructure, and play a major part in facilitating the access of information. There are, however, also a number of immaterial elements which, while perhaps less conspicuous than paper mills, xerox machines, electronic telephone exchanges or telecommunications satellites, are nevertheless equally important. One of these immaterial or software elements is the fluidity of scientific and technological communication. This concept of 'fluidity' of communication might be defined as the ease and speed of access to scientific and technological information and knowledge, both within a country and from abroad, and can perhaps best be illustrated with the help of an example taken from a scientific institution in India.

The National Chemical Laboratory in Poona is India's main central laboratory in chemistry. Out of the 350 professional scientists working there (most of them with doctorate level degrees), an average of six travel each year abroad for professional purposes (three on the Laboratory's budget and three on invitation from foreign institutes and professional societies). Statistically this means that the average scientist travels abroad once every sixty years, or to put things in more practical terms, that he has one chance in two of making one professional trip abroad during his working career.

It is of course impossible to measure how important foreign travel really is to the quality of a scientist's work: the rewards are not only professional but also psychological. In the case of India – and here the situation of the National Chemical Laboratory is fairly typical – it is extraordinarily difficult for the vast majority of scientists to travel abroad, and this lack of face-to-face communication with the outside scientific world is matched by an equally difficult access to foreign scientific publications, as even a brief look at the range of books and magazines in any institutional library will testify. This may be explained in a first instance by purely financial reasons: one should not forget that for an Indian scientist, the average cost of a two-week visit to the United States or Europe for instance is practically equivalent to his yearly salary. The problem however is not only financial, but reflects the more general attitudes of a society towards the outside world, its sense of its own importance and its pride in its long-standing traditions, and perhaps even a certain arrogance nurtured by centuries of brilliant achievements in the scientific field. Indian society, despite the openness due to its deeply democratic nature, is in some ways extraordinarily closed as far as communication with the outside world is concerned. While there may be a number of good reasons for this lack of communication, it clearly has a number of detrimental effects on the efficiency of the scientific and technological system, and more generally on the economy as a whole.

The patterns of scientific and technological communications with the outside world are also closely linked with a number of structural factors: large countries tend by nature to be rather more inward-looking than small countries, and countries which are scientific and technological leaders can rely much more heavily on their internal resources than countries which are merely followers or imitators. In the same way, rich countries can afford to spend more on communications than countries with very limited resources. Viewed in this perspective, the patterns of scientific and technological communication can be considered as a permanent feature in the intellectual landscape of a country, and for this very reason are probably rather difficult to change in a radical way.

What I would like to suggest here is that these patterns, despite their structural nature, can to some extent be modified. Increasing the fluidity of scientific and technological communication is in fact an important element in any government policy for information, and should receive the same attention as the building up of higher education institutions, the development of computer systems and telecommunications networks or the acquisition of foreign technology at the most favourable price.

The problem here does not lie in the technical or organisational means for improving the fluidity of communication, but rather in the way in which such expenditures are treated in the budgetary process of nations and institutions. In most research laboratories, or for that matter industrial firms and government departments, foreign travel, international telephone calls or subscriptions to major publications are considered as expenses which should by definition be minimised, since resources are limited. They could however also be viewed as investments, and should as such be subject to the same sort of cost benefit analyses as investments in scientific equipment or new buildings. Intuitively, one can sense that a $50 phone call to foreign colleague to solve a particular technical problem or to acquire an urgent item of information is intrinsically different from a similar expenditure on cleaning costs, for instance, in the same way that $10,000 spent on foreign scientific publications does not fit in the same category as $10,000 expenditure on a company car.

Modifications in accounting procedures alone are not of course sufficient to change attitudes or culture: in a sense, the lack of fluidity in a society's communications networks is very much a reflection of that society's values and traditions, and for this very reason is generally not perceived as a weakness or a failing of that society. It tends to be accepted as a normal feature which is seldom put into question, and it generally takes an outsider (i.e. a foreigner or a foreign-educated national) to realise how important the problem is, and how much more efficient the scientific and technological system could be, if communication with the outside world were not so difficult. What further reinforces this 'normality' of poor communications is the fact that societies tend to adapt themselves to the particular ecology of poor communication: the usual response to such a situation is generally not an effort to improve things, but rather a gradual reduction in the need for communication – a situation which of course further reduces the opportunities for communication. In the same way that people simply cease using the telephone when the quality of the network is too poor, societies reduce their needs for international communication when these needs cannot be satisfied under normal circumstances.

## Building-up a technological intelligence capability

Most developing countries are highly dependent on foreign technology supplied, often at great cost, by the industrialised countries of the OECD area. The weakness of their own research and development effort, coupled with the tremendous pressures to meet the needs and new aspirations of a rapidly growing population, suggest that in the foreseeable future, these countries will remain structurally dependent on foreign technology. If, as seems highly likely, their population doubles in the next thirty five years, and if we assume, rather conservatively, that real per capita income will double in the same period, this means that the total demand for new technology thirty-five years from now will be four times higher in real terms than it is today.

One of the critical problems facing these countries will therefore be to get access to foreign technology in an efficient and cost-conscious way. In a first analysis, the problem may appear to be one of information: where can the required technology be found? How do the costs and access conditions of technology compare with one another? How does one ensure that the full range of alternatives has been considered in any case of technology importation? Information in the widest sense is obviously a critical input into this process of technology identification, choice and transfer, but it is not sufficient. What is also required is a capability to make considered value judgements about the quality, appropriateness and price of foreign technology, and more generally an ability to interpret the broad trends in the international technological scene and identify promising new opportunities. In other terms, what is needed is a technological intelligence capability, and this capability, I would submit, is one of the centrally important instruments in any national information policy, and more generally, in a national development policy.

Like military intelligence, technological intelligence can be viewed as a three-stage process, and each of these stages raises a rather different operational and organisational problem. The first stage is that of intelligence *acquisition*; the second is the process of intelligence *evaluation*: and the third is the process of intelligence *utilisation*.

When considering the process of acquisition, it is important to keep in mind what might be called the '90-9-0.9 per cent principle': 90 per cent of the required information is open or freely available, 9 per cent (or thereabouts) is semi-open information, and less than 1 per cent is secret information which usually has to be obtained through illegal means. This rule-of-thumb about the accessibility of information is important in that it suggests where the technological acquisition effort of a developing country should concentrate, namely on the 99 per cent represented by open and semi-open information. This is not to say that the 1 per cent of secret information is unimportant: in many cases, this secret information can be critical, but an exclusive preoccupation with it may lead to a disastrous neglect of the open and semi-open information which is ultimately much more important.

The fact that by far the largest part of useful information is openly available does not, of course, mean that it is costless: it takes time, people and money to gather large amounts of information from open sources, to find out where this information can be found, to buy technical publications, to build up communications networks, and to file and cross-reference this information. It also requires creativity and imagination to transform raw data into meaningful information, and to build up a comprehensive picture of the technological system and its evolution from information which is incomplete, uncertain, unreliable or outdated.

The country which probably stands out as the best model in the field of acquisition of technological intelligence is Japan, and many developing countries could stand a lot to gain from a careful examination of the ways in which this country has systematically organised its acquisition effort – through innumerable technical missions in Europe and North America, through comprehensive surveys of the technical literature, and through a systematic pooling at the national level of all the items of technical information acquired by industrial corporations, trading companies and individual experts, as well as through the adroit exploitation of foreign experts. Partly as a result of this systematic acquisition effort, Japan has succeeded in getting access to foreign technology at relatively low costs (e.g. by identifying, and then buying, what might be called 'unfinished' technology, as opposed to the turn-key or 'product-in-hand' technology currently favoured by many newly industrialised countries), and has managed to overcome to a large extent two of the major obstacles to a Western-style industrial development, namely geographic distance and cultural isolation.

When stating that Japan is a good model, and perhaps the best model, to follow, I am not thinking so much of Japan today as of Japan in its early industrialisation phase (i.e. between the mid 1860s and the early 1930s) and during the 1950s and 1960s. This is the time when Japan was still a technological follower and was vitally dependent on foreign technology. Today, this systematic intelligence acquisition effort, built up over the decades, is still very active, but has lost some of its preeminence as a result of the simple fact that Japan has become the world leader in a number of industrial sectors (e.g. electronics, automobiles, steel-making, etc.), and is therefore rather less dependent on foreign technology than it was twenty or sixty years ago. These were also times when Japan was a poor country with very limited resources – a situation which is, here again, rather similar to that of many developing countries today.

The second important stage in the building up of a technological intelligence capability is the evaluation of intelligence: there is little point in gathering vast amounts of technological information if there is a no capacity to evaluate the quality of this information, assess its relevance to the needs of the country, and make the distinction between important and trivial information. Evaluation tends to be a rather qualitative activity, whereas acquisition is much more quantitative in nature, and it requires a high degree of technological sophistication to assess the importance of a technological development taking place abroad. In this perspective, the institutions which are probably most suited to carrying out this evaluation function are those which are already engaged in research and development in the country concerned, namely industrial laboratories and research centres in the public sector. From an organisational point of view, this evaluation function should probably be kept separate from the acquisition function – and can operate on a much more centralized basis than acquisition.

Because of its qualitative nature, evaluation is somewhat more difficult to manage than acquisition, and this is the stage where some of the most glaring failures take place: what retrospectively appear as intelligence failures are generally not failures in acquiring information, but errors in interpreting and evaluating this information. One typical illustration of this was the failure of the Swiss watchmaking industry to make a correct assessment of the threat presented by the advent of microelectronics – a failure which cost Swiss watchmakers over 50 per cent of the total number of jobs in this industry.[8]

As the interface between acquisition and utilisation, evaluation has a major role to play in directing and orienting the acquisition process: for countries with limited resources – and this is the case of most developing nations – the problem is not to acquire all possible technological information but only that which is truly important to the country and its major economic sectors. In this sense, assessment is not only a passive function (i.e. determining which information is useful and relevant, cross checking its quality and making this digested information accessible to potential users), but also has an active part to play as the initiator and selector of intelligence acquisition.

Acquiring and evaluating information, or more specifically technological intelligence, is of little use if this information cannot be used by its potential clients: the function of a technological intelligence effort is not simply to gather information and assess its quality, but to provide decision-makers in government, banks, industrial firms, research centres and other institutions with important elements of information that can be used in the design of policies and strategies, in negotiations with foreign partners, in establishing new development priorities and in determining the optimal allocation of financial resources.

In a sense, those people best qualified to determine what type of information is required are those who will ultimately be using it: people who really need a certain type of information are almost bound to use it once they have obtained it, either through their own efforts or through the efforts of an independent intelligence gathering unit. Close contacts between intelligence gatherers and evaluators on the one hand, and intelligence users on the other, is therefore of critical importance. However, close contacts alone are not sufficient: the function of intelligence gatherers and evaluators is not only to provide users with the information required by the latter, but also to make users aware of trends, innovations and opportunities which they are unaware of. Requests for information of this type are often difficult to make, since the users themselves are not aware that it exists, or that it may be important. In this perspective, the purpose of an acquisition or evaluation unit is not simply to provide high quality information at the requests of users, but to create a market, or develop a need, for the intelligence they have gathered, and which in their judgement, is important to the country's economic and technological future.

Without going here into the details of the ways in which a national technological intelligence effort could or should be organised by developing countries, it may be more important at this stage to suggest that countries contemplating a more active policy in the field of information start asking a number of basic questions about their technological intelligence activities. These questions can be summarised as follows:

(i) Does the country already have some sort of organised technological intelligence capability?
(ii) Do the potential users of technological intelligence feel a need for more or better intelligence about technological trends and developments abroad?
(iii) Approximately how large is the existing technological intelligence effort?
(iv) Is there any coherent national policy in the field of technological intelligence?
(v) Which are the weakest aspects of the country's present technological intelligence effort (acquisition? evaluation? utilisation?)
(vi) What would be the costs of building-up a basic technological intelligence capability?

(vii) Is technological intelligence acknowledged as an important component of a national information policy?
(viii) Are there any formal or informal training schemes for technological intelligence officers?

## The need for development intelligence

Technology is one of the most important forces shaping the future of developing and industrialised countries, and a national technological intelligence effort is not simply an interesting addition to the vast amount of work currently going on in the field of information, but a vitally important instrument of a more coherent development policy. It would, however, be a mistake to focus solely on technological intelligence. What developing countries require is more than technological intelligence: their future is also determined to a large extent by the strategies of large multinational corporations, by the lending policies of international development banks and other financial institutions, by changes in the relative prices of raw materials and industrial products, by the discovery of a new natural resources, by the development of new organisational tools and instruments, and more generally by the evolution of the world political situation.

Most countries have a fairly good political intelligence system, all military establishments have a rather sophisticated military intelligence capability, and there are few interior ministries which do not have an internal intelligence capability to monitor subversive activities on the home front or abroad. These intelligence activities are considered to be within the normal bound of governmental responsibilities, even though they tend to be carried out rather discreetly and under the cover of rather unassuming titles such as research, information, reporting or defence.

What I would like to suggest here is that, although most of the threats against which existing national intelligence activities are directed are very real, or believed to be real (which is practice amounts to the same thing), the most important challenges facing developing countries are neither military nor political, but social, economic and technological. In other terms, the real 'enemy' is not so much the insurgents from within or the adversaries abroad, but under-development in general. To use Stevan Dedijer's expression, development is a war,[9] and this war requires the mustering of all national efforts, including a national intelligence effort.

This intelligence effort directed towards development might be called, for want of another better term, 'development intelligence'. Technological intelligence is quite clearly one of its important components, but it may be useful here to outline a few others.

One of these is intelligence on multinational corporations, and more generally what might be termed industrial intelligence. Most developing countries have a number of foreign multinational corporations operating on their territory, and these firms often play a major role in a few key sectors such as mining, food processing, the production of consumer goods, and services (banking, management consulting, etc.). National attitudes towards such corporations tends to be rather ambivalent: their measurable contribution to employment, production, exports and technology transfer is often viewed as less important than the drawbacks, in the form of loss of control over economic decisions of key national importance, loss of national sovereignty, and the rather diffuse feeling that the host country is somehow being exploited by the sophisticated and footloose multinational corporation.

In most developing countries, the number of important foreign multinational corporations is very small: two or three in the case of small nations with a relatively undifferentiated economy, (e.g. Liberia), maybe 20 or 30 in a much large and more advanced developing country. Of these 20 or 30 firms, probably no more than 5 or 10 are truly important as far as employment, exports or investment resources are concerned. What is rather surprising to observe is the rather wide lack of knowledge within most host countries about the ways in which these major corporations operate, about their long-term plans, their development priorities, their psychology, their management styles or their attitudes towards technology transfer. This is to a large extent the result of the lack of a coherent intelligence effort directed at these corporations, and I would submit that such intelligence can be extremely important not only in the normal negotiation process with multinationals, but more generally in the continuing relationship between host country and foreign firm.

Another important area on which a development intelligence effort could focus is agricultural intelligence. In itself, this is nothing very new: the big grain trading companies, for instance, have developed what is probably the most sophisticated agricultural intelligence system in the world which monitors production, trade, technology and investment in all the major countries in the world, and they are among the major users of data provided by remote-sensing satellites.[10] The US Department of Agriculture is another of the big producers of agricultural intelligence about the United States as well as other countries. Developing countries, by contrast, tend to be very weak both in the acquisition and utilisation of intelligence: this comes out rather clearly from their very modest recourse to the services provided by the consulting firms which interpret satellite images, and on a more general level, by the lack of sophistication of their reporting systems for monitoring production, crop failures and transportation facilities. This is all the more important, given the major importance of agricultural production in the economy of these countries.

A third critical area for a development intelligence effort is what might be called organisational intelligence. Development depends not only on investment, education, manpower or technology, but also on new organisational forms and better policies. Although there is a wide amount of knowledge available throughout the world as to the most effective ways for organising a basic health care delivery service, for instance, or for that matter rural credit systems or agricultural extension services, it is surprising to observe how long it takes to transfer the experiences of one country or region to another, and how little is often known of highly successful experiments carried out in a neighbouring country. One of the functions of organisational intelligence should be to identify, compare and evaluate such experiments, and provide decision-makers with a much better knowledge about them, so as to allow for their application in the country concerned. Many experiments in the field of organisation are quite clearly country-specific and culture-specific, and therefore rather difficult to transfer into a totally different environment; a lot nevertheless is indeed transferable, and the problem is to find out about it. This is one of the functions of an organisational intelligence effort.

In conclusion, I would like to stress that if information is a critically important input into the development process, it is of little use if it is not well understood, badly evaluated and never used. A coherent development intelligence effort is not in itself sufficient to help solve all the problems facing developing countries, but it can play a useful part, and would deserve much more attention than it has received until now. It should be recognised as a legitimate and important instrument in the process of economic and social development, and of the tasks of researchers in the next few years could be to identify and evaluate some of the most interesting but little known efforts carried out by some countries in the field of development intelligence.

## Notes

1 The term telecommunications is used here interchangeably with the term telephone: investments in telephones account for over 90 per cent of all investments in point-to-point or two-way telecommunications.

2 These studies are summarised in W. Pierce and N. Jéquier, *Telecommunications and Development*, ITU and OECD, Geneva and Paris, 1982.

3 E. F. Parker, *Economic and Social Benefits of the REA Telephone Loan Programme*, ITU, Geneva, April 1982.

4 A. P. Hardy, *The Role of the Telephone in Economic Development*, ITU, Geneva, April 1982.

5 A. A. Kamal, *A Cost-Benefit Analysis of Rural Telephone Service in Egypt*, ITU, Geneva, April 1982.

6 M. Tyler, *The Impact of Telecommunications on the Performance of a Sample of Business Enterprises in Kenya*, ITU, Geneva, April 1982.
7 See for instance J. G. Bene, H. W. Beall and A. Côté, *The Tropical Forest-Overexploited and Underused: Research Priorities in Tropical Forestry*, International Development Research Centre, Ottawa, August 1976; and Forestry-Sector Policy Paper, The World Bank, Washington, DC, 1978.
8 This example is discussed in more detail in Nicolas Jéquier, *Intelligence as an Instrument of Development*, OECD Development Centre, Paris, June 1980 (mimeo)
9 Stevan Dedijer, *Social Engineering of Intelligence for Development*, OECD Development Centre, Paris, June 1980 (mimeo).
10 See for instance Gérard Blanc, *Grain Intelligence*, OECD Development Centre, Paris, June 1980, (mimeo).

# 9

# The Awakening Information Needs of the Developing Countries*

Björn V. Tell

## Introduction

Today there are many different approaches to providing technological information to a developing country. One way is to follow the traditions in the industrialised countries, and to set up a documentation centre. That centre will acquire all the major primary and secondary publications in science and technology, since these contain most of the information resulting from R & D in the industrialised countries. These publications may seem difficult to digest so the centre tries to hire qualified and language trained specialists to screen the material. Their task is to sift out information, translate it into abstracts to be presented in the abstract journal which is then disseminated to governmental and private enterprises in the hope that its contents should be applicable to the local situation. Often the centre is supported by a local printing shop where the equipment has been brought in through UNDP funds or other aid programmes, disregarding the fact that there are often already good printing facilities in the country.

Another way is to encourage multinational companies to set up affiliations in the country in the vain hope that the advanced technology they use will generate local innovations, and thus assure a technological leap for the country. The government ignores the fact that the multinationals seldom transfer R & D activities to subsidiaries abroad.[1] Many of the oil-rich countries try to buy turn-key factories including agreements about replacement equipment as well as the training of local personnel. Similar to that approach is the purchase of patents and know-how through consultants. In many cases these foreign consultants have not assessed the local techno-economic, environmental, and social factors in the country in which they are working.[2]

* This article first appeared in the *Journal of Information Science*, Vol. 1, No. 5, January 1980.

The usual approach is to set up a centre intended to assist the country. This institutional approach has been heavily stressed by Unesco, especially in the UNISIST-programme.[3] A number of institutions like INSDOC, PANSDOC, CORSTIC, etc. have been set up by Unesco; or 'replicas' like ISTIC in China, CONICIT in Venezuela, and JCTND in Yugoslavia, have been set up by national governments. However, in many cases these institutions seem not to have progressed as hoped.

First, the status of the information workers in these centralised documentation centres is often very low. The professional librarians are regarded as mere custodians of books on shelves. Second, much of the work in the centres is governed by complicated international cataloguing and abstracting rules that result in low productivity and overall low efficiency.

In the two other cases mentioned above it is assumed that the import of hardware of a variety of kinds will give the country the missing technology. In both cases the policy of the governments is responsible for that state of affairs, because their officials are constrained to work in an unimaginative way relying upon old concepts and traditions to find an answer which transpires not to work in the way in which was expected.

Another reason for the backwardness relates to the thin layer of educated people with knowledge of languages in central administration. Most developing countries have serious inadequacies when it comes to knowledge of languages especially in the mastering of a few of the world languages like English, French or Russian, languages in which the world's scientific and technological knowledge is exposed in the primary and secondary journal literature. There is an acute shortage of technical writers in these countries. This limited capacity to tap international information sources is often combined with a limited capacity for local innovation because the poorly developed information transfer mechanism does not work.

So, even if we have now come across some important factors which need more consideration, we are still far from answering the original question of what information is at the disposal of governments, and why do they not design effective knowledge networks to take advantage of the world's knowledge? During recent years there have been strong voices both in the United Nations and in Unesco for the establishment of a new international order which will make it possible to eliminate the widening gap between the developed and developing countries.[4]

## Power of new information technology

In recent years the new information technology has completely changed the traditional way of gathering, retrieving and disseminating scientific and technical information in the industrialised countries. The advent of computers, electronic communications networks, and huge databanks have contributed to closing the information gap which was earlier thought to be unavoidable because of information overflow. In many Western countries, using terminals in libraries and information centres is a daily routine. The tapping of a database in Italy or California is now performed in the same way as linking up with a local computer in the same building or on the same campus. Online searches in the world are up to more than two million terminal hours a year to the 500 databases which are publicly accessible, and the immediate yearly growth of use is expected to be 30 per cent.

It has been said by leading personalities of the non-governmental organisations that the new technology and the use of computers for information retrieval and the like is not yet suitable for the developing countries. Instead, it is claimed that those countries should build up a meticulous infrastructure of service organisations by emulating the traditional libraries and documentation centres of the industrialised countries. Such a task is commendable but does not take into account the time factor, nor the shortage of personnel, premises and material.

A closer examination shows that such a view is untenable. Demonstrations of online searches made by the European Space Agency and Unesco in countries such as Morocco, India and Kenya, show an interest and a growing demand for online retrieval services among the developing countries. Demonstrations by commercial organisations such as Lockheed and Systems Development Corporation in various parts of the world have increased the appetite for utilising the accumulated knowledge in databanks. It may also be noted that TYMNET is riding on the wave by setting up nodal facilities in places like Singapore and Manila in order to facilitate the network accessibility to the developing countries of Southeast Asia.

The demands of the developing countries cannot be ignored any longer when it comes to the development of modern information and communications systems. Instead, the needs expressed by these countries should receive priority attention by Unesco, UNIDO and other organisations. It is therefore necessary to improve Unesco's activities within the UNISIST programme so that mechanisms are set up in the developing countries to connect them with the information networks. So, emphasis must be placed on the structure of such mechanisms in contrast to traditional services. That is where computers and communication technologies come in to provide a basis for a new information infrastructure which will permit the transition from 'present, essentially *static* information as represented in books, journal articles and other printed matter to *dynamic* information in electronic form'.[5]

## A new model

Recently, one developing country has taken a forward-looking decision to create a mechanism which may be taken as a model structure for other countries under development. Starting off with traditional services like libraries, archives and documentation centres, one ministry has coordinated all such activities together with the computer and communication technologies under one head. In other words, the Ministry has set up an 'intelligence' function within its domain to cater for technology transfer. The Ministry which is responsible for energy and mining has an essential stake in the nation's economy, and thus responsibilities for trade pact negotiations, foreign aid negotiations, licence purchasing, major systems purchases, hiring of consultants and their utilisation, national equity negotiations, catastrophies and disaster alleviation agreements, state and privately owned enterprises, foreign technology transfer, domestic technology transfer, joint ventures and cooperative endeavours technology transfer.

Within the hierarchical structure of the Ministry (which also embraces nationalised industry), several information analysis centres and service units have been set up at divisional and sectional levels. Together with the central library and the computer service centre they form an internal information network primarily to support the information system requirements of the Ministry. For instance, within the Directorate of Geology, four such information units have been set up; two under the Directorate of Petroleum Economy, and one in each of the other seven directorates. These units are acting as branches of the 'information management and service centre' which instead, and in accord with tradition, is still called the Central Library of the Ministry. By the development of this in-house information resource it is hoped that the information needs of the Ministry, related agencies and the industry will be met.

The Ministry has thus developed an effective institutional mechanism and expertise to ensure the performance of various functions for the assessment and selection of appropriate technology for application and adoption in one of the most productive sectors of the economy. Its computer centre plays an important role. A series of programmes have been implemented which makes it possible on various topographical, geological, economic, and other maps to locate the interesting geographical and geotechnical sites. Water wells, as well as oil wells, are mapped out and described in detail. The analysis centre for geological and mining planning has processed and codified more than 25,000 maps, charts, sectional drawings and bore cores.

One information analysis centre for technical reports on economic geology and mining has analysed 2,600 technical reports and coded them for the computer, using TEXT-PACK for retrieval purpose. In the field of oil economy another specialised information analysis centre has acquired 10,000 information items, and another centre collects clippings from daily newspapers all over the world that are put into the computer. Also, all oil-wells documented in 26,000 technical reports, are in the database. All interesting geophysical data throughout the country have, on an experimental basis, been put into a database. For a tropical country the water resources play an important role, and another analysis centre is collecting and analysing the information available about subsoil water and wells in order to make it available in machine readable form for the use of decision makers in the Ministry for Environment concerned with renewable natural resources.[6]

The geotechnical database also has applications and users outside the Ministry, especially for urbanisation projects and for construction and civil engineering works. In other words, the Ministry has created an 'intelligence' function, by which is meant the capacity to obtain, process and use information so that it can pursue and achieve its goals, whether it has defined them itself, or whether these goals have been assigned to it by a government plan. The Ministry has to apply the secrecy necessary for the functioning of affairs, but it also opens its services to other ministries, and in certain cases to public and private enterprises.

In the nationalised petroleum industry a special centre for technical information has been set up in close collaboration with the Ministry, in order to identify information sources of interest for the development of the industry. That centre makes heavy use of international databanks accessible online. The Government has allowed the centre to share a direct, rented line to the United States that originally was set up for other purposes.[7]

This case study of the development of an indigenous information function presents a lesson for the traditionalists in the international non-governmental organisations who believe that technological transfer mechanisms must rely on the establishment of traditional libraries and archives on the Western pattern, while at the same time they condemn the use of computers and terminals as being premature attempts of modernisation. Such views can only increase the information gap and the lead time associated with the introduction of new information technology in the developing countries.

As mentioned earlier, the layer of qualified people with knowledge of languages is very thin in the developing countries, and such people are often employed in the ministries and the central government agencies. The setting up of a knowledge network in these quarters may provide the possibility for a developing country to take advantage of the present information transparency in the industrialised countries by making use of terminals for accessing the worldwide information databanks. Even if the huge databanks may seem difficult to exploit, an online dialogue will still help the developing country in the enormous amount of work of sifting through the primary and secondary literature. Present findings show that almost all information necessary for intelligence action and national planning is openly available, and obtaining such information today is not as costly as many decision makers in the developing countries may believe. But it needs rethinking and reorientation towards the use of modern information technology.[8]

## Conclusions

The infrastructure of information services suitable for developing countries is a matter for international concern. The timescale needed to build a structure of traditional libraries and information centres, however commendable and attractive that idea may seem to Westerners, is much too long in most cases due to the acute shortage of resources, premises, trained staff, etc.

The concentration of qualified personnel in the ministries and the nationalised industries that is common in the developing countries should be the main focus of attention. Therefore, a different approach could be argued, stemming from the enthusiasm with which many developing countries have taken to online systems when demonstrated there, and from a few new information centres which could be taken as models for a different structure. What seems to be needed is the setting up of ministerial 'societal intelligence' functions integrating libraries, documentation centres and information analysis centres, interwoven into ministerial information networks.

One possible approach would be to establish a few national or regional centres in order to demonstrate the feasibility of this different approach. Such centres which could develop into 'centres of excellence' could initially be located in a country which could provide the necessary support in terms of equipment, maintenance, supplies, postal and telephone communications, etc., and to which it would be possible to attract the highly qualified and experienced staff who would be needed in the systems analysis and applications programming areas.

The databases loaded at such a centre for online appropriate information for the countries in that region. Gaps in the databases would be identified and initiatives stimulated to fill these gaps. The centre could have the dual function of providing an operational service to the neighbouring countries whilst simultaneously serving as a training centre for personnel from those countries who would later return to support local services.

For the host country it would be attractive to eventually take over such a centre. Therefore, the sharing of its costs might show up to be a good investment. The annual costs for such a centre of excellence might run up to over a million dollars. Compared with what a country is spending on military intelligence that sum would seem to be of minor importance for the establishment of a 'social or societal intelligence' function which would have a more important stake in the economy than the labour intensive but low productive military 'intelligence function' could ever contribute. Some of the costs might even be recovered by means of a subsidised charging system for the users of the services.

## Notes

1 P. B. Evans, 'Foreign investment and industrial transformation: A Brazilian case study', *Journal of Development Economics*, 3 (1976) 119–140.
2 O. Hawrylyshyn, 'Capital-intensity biases in development country technology choice', *Journal of Development Economics*, 5 (1978) 215–236.
3 *Guidelines on the planning of national scientific and technical information systems* (Unesco, Paris, 1975).
4 *Moving towards change* (Unesco, Paris, 1976) 9.
5 H. P. Gassmann, 'Data networks: A new information infrastructure'. *OECD Observer*, 95 (1978) 11.
6 D. Zozaya R., Red Interna de informacíon técnica del Ministerio de Energia y Minas, Modelo para la red nacional de información, Proc. 10° Congreso Venezolano de Ingeniera, Arquitectura y Profesiones Afines, Puerto Ordaz (1979) 35 p.
7 A. S. De Gutierrez, C.I.T. Un centro de Informacion tecnica petrolera, Caracas, INTEVEP (1975).
8 B. V. Tell, Information as a resource – a cost estimate, *Int. Forum Inf.* Doc. 4 (1979) 36.

# 10

# Some Reflections on Economic Information Gathering and Processing within the Framework of the North-South Conflict

Miguel Wionczek

## Introduction

Having been engaged for some twenty years in the activities related to gathering, processing and diffusing economic and technological information for policy-making purposes in Latin America, I wish to consider some practical experiences of building information handling capacity in the developing world. My story will deal more with the preconditions for the success of such exercise than with detailed ways of gathering and processing what falls generally under the rubric of 'economic intelligence'. My own experiences strongly suggest that the obstacles and difficulties using specialised information to maximum advantage are less technical than social, political, and professional, and that these obstacles spring largely from the nature of underdevelopment itself.

Two practical examples may help to illustrate the problem. A distinguished Latin American historian offered the proposition that most scientists including those working on applied economics required only two to three years of intensive learning to reach an outstanding level of expertise in a given field. He maintained that, given the availability of a reasonably good educational background, moderate intelligence and intellectual curiosity, it is easier to master a new discipline in a few years than by studying it over a lifetime. A lifetime's dedication to a single subject tends to burden one's mind with many biases and prejudices. Moreover, such patient dedication provides us with a great deal of useless and irrelevant information which makes it more difficult to direct attention to the key issues of enquiry, particularly if the subject is multidisciplinary and the results of enquiry are to be applied to policy-making. There is always the risk of becoming fascinated with the method rather than with the substance and the purpose of enquiry. For some evidence in this respect a perusal of any issue of *American Economic Review* is more than adequate to illustrate the point.

The second example is based on the concrete aspects of information gathering for policy purposes. In the early 1970s, when I was charged with science and technology policy problems in Mexico, a retired Israeli general was passing through the capital. His job was to gather and analyse world-wide information on advancements in military technology. My interest in meeting him had nothing to do with the field of his concern since I deeply abhor anything related to war. I was aware, however, that the Israeli establishment was extremely successful in building up its own military technology capability. Consequently, I was keen to find out how mechanisms of gathering relevant overt technological information worked in that country.

The story which was recounted by the general was impressive because of its simplicity. His information gathering operation consisted of a score of well-trained experts in military technologies, with the working knowledge – one must add – of several foreign languages. They had constant and speedy access to not more than one hundred political and technical journals dealing with military technology matters world-wide (including commercial publications of the arms-producing and -selling firms). While the annual output of printed material on military technology subjects must weigh tons, the general considered that over 90 per cent of the material (including most books) was irrelevant for his exercise, being out-of-date, repetitive, or of poor technical quality. Collecting information on the factors of greatest relevance in a field, characterised by an extremely rapid rate of advancement, is the first condition of success. The second is the knowledge of the state of art. The third is the ability to integrate pieces of information into coherent and usable units. Finally, an effective infrastructure is required, composed of organisations able to make use of information gathered with policy problems in mind. Policy requirements must be well defined in order for this process to operate maximally.

These key assumptions are worthy of closer scrutiny in the context of the conditions prevailing in developing countries, beginning with the following propositions: first, the world is passing through an accelerated scientific and technological revolution of previously unknown magnitude; second, information about the process and implications of this revolution is distributed with great asymmetry; third, information of varied kinds is of pivotal importance to global political and economic relations, particularly North-South negotiations; and fourth, greater access to and control of information helps the North to keep the South in a state of considerable political, social, economic and technological dependence. Something ought to be done urgently to improve the conditions of information gathering, processing and user capability in the South. If some effort were made, asymmetries of all types would be considerably reduced and prospects of negotiating the existing conflicts would improve. While in principle I have no quarrel with the proposition that information is pivotal in international relations, I have the most serious doubts whether it applies to all types of information.

## The scientific and technical revolution

That the world is passing through an accelerated scientific and technological revolution of magnitude and speed previously unknown is a highly debatable issue on a number of grounds. First, no way of measuring the pace of advancement has been found as yet except in exact sciences and some high technology sectors, where the measurements are still very primitive and controversial. In exact and natural sciences (and more and more in social sciences as well), a particular yardstick is to count the printed output and the frequency of its use by the scientific community through citations. In high technology sectors the progress is measured by the speed with which scientific discoveries are translated into new technologies, and these in turn are incorporated into the production of goods and services. But both measurement techniques are open to discussion.

In respect of the high technology advancement, measured by the volume of product sales in which such high technology has been successfully incorporated, doubts as to whether this reflects technology information advance are even more substantial. The case of the worldwide growth in the information industry is perhaps illustrative here. Anyone cognizant with the state of information about important economic and financial phenomena in most of Latin America, for example, must have little doubt that the mass invasion by the computer sciences is largely a result of fashion and selling techniques, not a reflection of information explosion. Literature on the need to computerise grows, however, alongside the sale of hardware and software; this in turn, may serve a reflection of the measurement of the pace of technological revolution.

## Information and policy

My particular concern in recent years has been the international energy field – a subject of vital policy importance as well as particular fascination to thousands of people all over the world. Hundreds of books in all possible languages and locations appear annually on the world energy subject. The number of new scientific and pseudo-scientific journals in the field has been increasing even faster than the output of books. If any particularly curious and mobile individual, devoid of any financial constraints, would like to attend all the congresses, international conferences and workshops on the energy subjects, he would need some five years of time every year! After strategic theories and military research, the energy 'research' business seems to have emerged as the biggest scientific and technological business in the world.

In reality, it is not difficult to know what is going on in this particular research field. It is sufficient to attend a dozen well-chosen international meetings annually, acquire not more than 30 also well-chosen books, and subscribe to perhaps twenty professional journals and several international financial newspapers. It is more or less *all* that is needed to be fairly well informed about political, economic, financial and technological developments in the energy field. An informal world-wide network with the people in the know will be, of course, of considerable help. It will be available, however, only to people who have mastered the state of art and count with information about sources of information.

Information is pivotal to international political, economic and other relations, once it is organised functionally to serve well defined purposes, preferably related to the positive goals of optimising economic and social welfare. The dictum that 'information is power' is not correct since only usable information may become a source of power. The accumulation of information may lead, according to a given situation, either to easier conflict, resolution or to further conflict. Information is not neutral, and its usefulness certainly depends more on the precision of its application than on its availability. While availability of all types of information in the South may be useful for improving bargaining capacity with the North, its usability really depends more on building up the capability to exploit information potential than on the volume of information available. The collection of information must be selective and functional.

In the energy field, a meaningful and relevant national energy information service is within the capacity of most middle-size semi-developed countries of the South. In this particular sector of vital policy importance, the obstacles to having access to information about the shape of things taking place and likely to occur within reasonably short time span are not of a technical nature. They are largely to do with political management at the macro level and of bureaucratic capacity at the micro-level.

The political management capacity required in a developing country to build up the information gathering, processing and user capacity, especially in international matters, is extremely low for a number of reasons. At the highest political level, centralism, accentuated by the prevalence of parochial nationalism, is a major obstacle to the requirement of national economic intelligence services of international scope, in which informal exchange of information is as important as the access to data banks or the range of computer services.

A second obstacle may arise because of the absence of highly trained personnel in specific international economic issues, and the low esteem in which such people, when available, may be held by the governing class. Any politician is naturally wary of experts and professionals. Consequently, most social economic intelligence tasks are performed mainly by mediocre bureaucrats.

The third obstacle may have its origin in the political and administrative instability from which most developing countries suffer. In relatively democratic societies the political process often consists of a high degree of professional mobility which makes responsible and efficient functioning of the agencies charged with technical tasks almost impossible. Gathering and processing international economic information is, in fact, a long-term exercise. In authoritarian societies the conditions are even worse, because technical capacity may be diverted into other uses.

The fourth difficulty may be found in the bureaucratic competition between the holders of some specific segments of information (whose number grows constantly with the expansion of the public administration) and the public enterprise sector. While it is quite obvious that in the private sector, because of the prevailing atmosphere of commercial and industrial secrecy, the diffusion of information is not appreciated at all; the non-diffusion of information is also quite prevalent in the public sectors of the developing countries. Information, most often worthless because of its paucity and unreliability, is considered a source of power or a step towards the achievement of power.

## Conclusion

In the business of economic intelligence gathering, most of the obstacles considered replicate themselves on a smaller scale. This situation strongly suggests that the task of collecting, processing and diffusing economic information (because of its possible uses for domestic political purposes) is both highly politicised and problematic.

Does it mean that these vital activities are impossible and that the present asymmetry between the availability of economic information for negotiating purposes between the North and the South must continue to persist? The clear understanding of difficulties and obstacles might help to improve the situation. Some practical suggestions might be found in the following:

(i) Collecting, analysing and diffusing international economic information may be helpful to external economic policy formation, as a political as well as a technical exercise involving both forms of high-level skill.

(ii) No such exercise is possible if it is organised on an *ad hoc* or emergency basis, or if the users of information are unable to be explicit in their demands.

(iii) Since one of the basic conditions of success is broadly based knowledge of the sources of relevant information, economic intelligence for policy-making purposes must dedicate large part of its efforts to the art of gathering 'information about information'. It ought to give particular emphasis to detecting the biases of the information sources, with clear understanding that totally unbiased or 'neutral' information is not a real construct.

(iv) The purpose of the exercise is not only to acquire the relevant information but to have capacity to integrate it into policy-oriented packages.

(v) The people who engage in economic information gathering and processing must be more than just good professionals in their respective fields. They ought also to be intelligent and curious, to have a multidisciplinary background, be well-read and well-travelled, problem-oriented, future-looking and risk-taking.

Economic intelligence units ought to be small and relatively agile in their dealings with the administration, based perhaps at universities or in some other autonomous relationship to governments. With relatively little resource outlay, their contribution to policy could be substantial. In his article in this volume Björn Tell considers another type of policy intelligence unit, its location and principal activities. A useful exercise for improving the information handling capability in many developing countries may be to consider carefully the salient characteristics of these small units – many of which are based in Latin America – including their possible serviceability to other national policy environments. Among the most important policy areas in which such intelligence units are most needed for developing countries are: finance, food commodities, minerals, energy alternatives, sources of technology and forecasting. Such an initiative would greatly assist a rationalisation of the international consultancy market and give much more power over resources and opportunities to local professionals and institutions. It is one of the most important short-term improvements that should be made in the gap in information access which persists between North and South.

# Index